Fundamentals of Agricultural Statistics

NIPA® GENX ELECTRONIC RESOURCES & SOLUTIONS P. LTD.
New Delhi-110 034

Fundamentals of Agricultural Statistics

Abhiram Dash
Assistant Professor (Agricultural Statistics)
College of Agriculture, Chiplima
Odisha University of Agriculture and Technology
Bhubaneswar, Odisha

NIPA® GENX ELECTRONIC RESOURCES & SOLUTIONS P. LTD.
New Delhi-110 034

NIPA® GENX ELECTRONIC
RESOURCES & SOLUTIONS P. LTD.

101,103, Vikas Surya Plaza, CU Block
L.S.C. Market, Pitam Pura, New Delhi-110 034
Ph : +91 11 27341616, 27341717, 27341718
E-mail: newindiapublishingagency@gmail.com
www: www.nipabooks.com

For customer assistance, please contact
Phone: + 91-11-27 34 17 17
Fax: + 91-11- 27 34 16 16
E-Mail: feedbacks@nipabooks.com

ISBN: 978-81-19103-65-2

Composed and Designed by NIPA®.

Preface

This book is designed especially for under Graduate students of Agriculture and Allied Sciences and has been prepared in consonance with the syllabus recommended by the Fifth Deans Committee. The book would also be immensely helpful to agriculture graduates preparing for JRF and other competitive examinations. The concepts in the book have been analysed and presented in a simple and precise manner to the comprehension of the students. I hope, my sincere endeavour would cater to the needs of the students and render them immense benefit.

I thank M/s New India Publishing Agency, New Delhi for taking up the publication of the book in neat, flawless and presentable form.

I owe indebtedness, pay profound ovation and dedicate this little effort of mine to my departed parents who always prodded me for academic excellence and dissemination of knowledge.

Sri Gundicha **Dr. Abhiram Dash**

04 July 2019

Contents

Chapter 1

Introduction to Statistics

The word 'statistics' has been derived from the Latin Word 'Status' which means 'Political State'. The word statistics has been coined by Gottfried Aschenwall, a German Political Scientist. Ronald Aylmer Fisher is regarded as father of modern statistics. The word statistics has meaning both in plural and singular sense. In plural sense, it means quantitative/qualitative figures. For example, the area under food grains in different states of India, the production of guava in different states of India, germination percentage of seeds, etc.. In singular sense, it means as a subject i.e., method of dealing with the quantitative/qualitative information.

Statistics is defined as scientific method which deals with collection, organization presentation, analysis and interpretation of data. It is included under both science and arts. Statistics is used as a tool to obtain knowledge on other subjects of science.

Uses of statistics

Statistics has many uses in the areas such as economics, agriculture, business, medical science, engineering, industries, biology, psychology, planning & budgeting by the government.

Statistical methods are used now-a-days in all branches of agriculture for efficient planning of experiments and for interpreting experimental data.

In agricultural sciences, statistics can be used for the following purposes;

(i) To analyse the effect of fertilizer application on yield of crops, vegetables, fruits etc.

(ii) To study the chance of pest incidence and estimate the effects of pest incidence on yield of crops, vegetables, fruits.

(iii) To forecast the future production of crops on basis of past performance.

(iv) To test the performance of new varieties and new technologies by comparative analysis.

On basis of application of statistics, it can be categorized as:-

(i) Descriptive Statistics: The branch of statistics dealing with summarization of data (i.e., organizing and presentation of data) is known as descriptive statistics.

(ii) Inferential Statistics: It deals with analysis and interpretation of summarized data.

(iii) Applied Statistics: It consists of application of statistical methods and techniques to the problems in different fields such as agriculture or any other allied subjects.

Limitations of statistics

Statistics also have some limitations. They are the following:

1. Application of statistical theories can be done only when there is variability in the experimental materials/objects under study.
2. Statistics deals with aggregates or groups and not with individual objects.
3. The laws of statistics are true only on an average in the long run. For example, when we say that the chance of pest incidence in a particular fruit crop grown in a field is 20%, it does not mean that in every batch of 100 plants exactly 20 will be affected by the pest. This proportion will hold good only in a long sequence of observations.
4. Statistics is liable to be misused when used by inexperienced personnel. The misuse of statistics can lead to wrong conclusions which could misguide the person using it for some purpose.

Data and Information

Without data, statistics has no existence. Data is the raw material for any statistical analysis. Data is defined as collection of quantitative figures or qualitative facts obtained by measuring or counting in case of quantitative figures and by observing in case of qualitative facts.

Data is always in raw form. Data when arranged in a meaningful way becomes information.

Data is a plural term. Singular form of data is datum. The word 'information' can be used as both singular and plural term.

Classification of data. Data can be classified on different basis

1. On the basis of sources of collection, data are of two types: Primary data and Secondary data

Primary Data	Secondary Data
These are collected directly from the field of study.	They are not collected directly from the field of study. They are collected from some sources.

Primary Data	Secondary Data
Different methods of collection of primary data are: i. Direct oral interview ii. Collected through mailed questionnaires. iii. Collected through schedules by the enumerators iv. Old records	Secondary data are collected from i. Published sources such as, Agricultural Statistics at a Glance, Economic Survey, Odisha Agricultural Statistics, etc. ii. Unpublished sources such as govt. records from Government organisation, records from private agencies etc…
Data collection by this method is laborious, time-taking and costlier.	Data collected by this method involves less labour, less time and less cost
Data collected is highly reliable.	Data collected is less reliable.
Example - census data, number of tillers per plant in rice crop grown in the field, data collected by investigators etc…	Example- published thesis, project reports, monthly abstracts of different publications etc…

Basing on the nature of the characteristics observed, data are classified as

i. Quantitative data
ii. Qualitative data

Quantitative Data: Numerical data which are obtained either by counting or by measuring are called Quantitative data. Quantitative data are based on Quantitative Characteristics. The Quantitative Characteristics are called as **variables**. For example, area under rice crop, yield of rice crop, production of food grains in India, number of grains per panicle, length of panicle, age of trees, number of fruits in a tree, fruit size, etc.

Qualitative Data: Data having only literal meaning and no numerical value but may be assigned some numerical values, such as ranks are called Qualitative data. Qualitative data are based on Qualitative Characteristics. The Qualitative Characteristics are called as are called attributes. For example, gender, religion, marital status, shape, size & colour of seeds, class of farmers (marginal, small, medium, large), educational qualification, etc.

Variable: The quantitative characteristics whose values are obtained either by counting or by measuring for statistical study are called variables.

Variable may be:
(i) discrete
(ii) continuous.

(i) **Discrete Variable:** The variable where fractional measurements are not possible are called discrete variables. Usually discrete variables are obtained by counting. discrete variables can take finite values within the prescribed range. For example, the number of grains per panicle of paddy, number of tillers per plant, etc.

Continuous Variable: The variable where fractional measurements are possible are called continuous variables. Usually continuous variables are obtained by measuring. Continuous variables can take any value within the prescribed range.

Examples of continuous variable are plant height, seed weight, fertilizer dose etc.

Methods of collection of data : There are usually four methods of data collection:

1. **Direct Oral Interview:** Data collected directly from the informant by face to face contact with the informant. This method is suitable where intensive study of limited field is required.

   Merits of the method: (a) The interviewer can adjust the language of communication as per the requirement. (b) Chance of non-response decreases as the interviewer will get a chance to convince and motivate the respondents who are unwilling to give furnish data.

   Demerits of the method:(a) This process involves high cost. (b) It is more time consuming and laborious.

2. **Mailed questionnaire method:** A set of questions related to the study are sent to the informants by post who fills them and sent it back. A covering letter must be attached with the questionnaire which contains the purpose and benefits of the survey for which the questionnaire is prepared. This method is suitable where less details study of wider field is required.

   Merits of the method: (a) Involves less cost. (b) Less time consuming and less laborious as compared to direct oral interview method of collecting data.

   Demerits of the method: (a) No scope for changing the language of communication, if required which may lead to non-response from the informant. (b) Chance of non-response can also increase if the informants are not willing to furnish the desired data as per the questionnaire sent to them.

3. **Schedules sent through enumerators:** A set of questions related to the study (known as schedule) is prepared by the investigator and carried to the informants by the enumerator who fills in the schedule after collecting data from the informants. This is the most widely accepted method. This method overcomes the demerits of the direct personal interview method and mailed questionnaire method.

4. Old records: It involves collection of data from books, records, diaries, etc. which are relevant to the study and then the organizing the collected data as per the requirement of the investigator.

Organization of data: It includes editing & classification of data. Editing of data means to scrutinize the data to detect possible errors & irregularities. It is done to make the data complete, consistent & accurate.

Classification of data is done to facilitate the presentation and analysis of the data. The classification of data can be on several basis as per the requirement. Some of the usual type of classification of data are the following:

1. Temporal classification (on basis of time): e.g., Production of food grains in India from 1980–2016.
2. Spatial classification (on basis of region):e.g., Production of wheat in different states of India in the year 2015.
3. Classification by attributes or qualitative classification. Attribute is defined as qualitative characteristics having no numerical expression. e.g.-religion, gender, qualification, colour of flower, shape of fruits, etc.
4. Classification by variables or quantitative classification(on basis of quantitative characteristics). Variable is defined as the quantitative characteristics having numerical expression. E.g., height of plants, no. of fruits per plant, no. of grains per panicle.

Chapter 2

Frequency Distribution

Frequency distribution table is a statistical table which shows the sets of all distinct values of the variables arranged in order of their magnitude either individually or in groups with their corresponding frequencies side by side.

Need for frequency distribution table

1. To condense the data
2. To provides quick reference for the entire data
3. To facilitate quick & easy statistical analysis of data
4. To reveal the pattern within the data

On the basis of types of data, there are two types of frequency distribution

1. **Ungrouped frequency distribution:** An arrangement of raw data individually along with their respective frequencies is called ungrouped frequency distribution. This type of frequency distribution is prepared when total no. of observations (N) are large but no. of different observations (n) are less.
2. **Grouped frequency distribution:** An arrangement of raw data by putting the whole range of observations into a no. of smaller groups or classes, showing the respective class frequencies against the classes, is known as grouped frequency distribution. This type of frequency distribution is prepared when total no. of observations (N) and no. of different observations (n) are both large.

But when total no. of observations are less, there is no need of frequency distribution. In this case, the data series is known as individual data series or simple data series.

Terms related to frequency distribution

1. **Class interval:** The groups into which the values of a variable are assigned is called as class interval. E.g., 20–30, 45–55 are class intervals.
2. **Class limits:** The two end values or the maximum and minimum values of a class interval are called class limits. The maximum value of a class interval is called Upper Class Limit and the minimum value is called

Lower Class Limit. E.g., For the class interval 35–45, 35 and 45 are the class limits. 35 is the lower class limit and 45 is the upper class limit.

3. **Class mark or Mid-Value** of a class interval – It is the average of the lower class limit and upper class limit. It is considered as the representative value of the class interval. The class mark or mid-value of the class 30−40 is $(30 + 40)/2 = 35$.
4. **Class size:** It is the difference between the lower and upper class limit of a class interval. Also called as class length or class width. The class size of the class 35 − 45 is 10.
5. **Frequency:** Number of times a particular observation is repeated is called its frequency.
6. **Percentage frequency:** (Frequency/Total frequency) × 100
7. **Relative frequency:** Frequency/Total Frequency
8. **Frequency density:** Frequency/class size
9. **Class frequency:** The number of values of a variable falling within the limits of a class interval is called class frequency of that class.

Grouped frequency distribution is of two types:

1. **Exclusive type:** The frequency distribution in which the upper class limit of the classes is excluded from the respective classes is called Exclusive Frequency Distribution. This type of frequency distribution is mainly for continuous variables like age of tree, yield of crop, plant height, seed weight, etc.

   The classes whose upper limits are excluded from that class are called exclusive classes. In such classes the values of the observations which are greater than or equal to lower limit of the class and less than the upper limit of the class are included i.e., $LCL \leq x < UCL$.

   Example: The class intervals such as 20–30. 30–40. 40–50 are exclusive classes.

   Here in the first class the upper class limit 30 is excluded, in the second class the upper class limit 40 is excluded and so on.
2. **Inclusive type:** The frequency distribution in which the upper class limit and lower class limit, both are included in the class are called inclusive frequency distribution. This type of frequency distribution is mainly for discrete variables like number of tillers per plant, number of grains per panicle, etc.

   The classes whose both upper and lower limits are included are called inclusive classes. Here, $LCL \leq x \leq UCL$, where x is the value of the observation.

Example: The class intervals such as 20–29. 30–39. 40–49 are inclusive classes.

Here in the first class both the class limits 20 and 29 are included, in the second class both 30 and 39 are included and so on.

Tally marks: These are the lines drawn to represent the frequencies of observation or class frequencies of class interval.

Class boundaries or true class limits: When the inclusive classes are converted to exclusive classes, the class limits of these classes are called as class boundaries or true class limits.

Inclusive classes are converted to exclusive classes in the following manner:-

True lower limit or lower class boundary = Lower limit – $d/2$

True upper limit or upper class boundary = Upper limit + $d/2$

d = Lower limit of any class – Upper limit of the preceding class.

The frequencies of the classes will remain unchanged.

Example: 30–39, 40–49, 50–59 are inclusive classes.

Here,

d = Lower limit of any class interval – upper limit of the preceding class interval

= 40 – 39 = 1

The exclusive classes are obtained by subtracting 0.5 from their lower class limits and adding 0.5 to their upper class limits. Thus the exclusive classes are 29.5–39.5, 39.5–49.5, 49.5–59.5. The class limits of the exclusive classes are called as true class limits or class boundaries.

Cummulative frequency (c.f.)–Cummulative frequency (c.f). of a class interval is the number of observations which are less than true upper class limit of the class interval (upward c.f.) or number of observations which are more than or equal to true lower limit of the class interval (downward c.f.).

In the exclusive Frequency Distribution given below, the upward and downward c.f. are obtained in the following manner:

Class interval	0–10	10–20	20–30	30–40	40–50
Frequency (f_i)	7	8	10	6	4
Upward cumulative frequency (u.c.f.)	7	15	25	31	35
Downward cumulative frequency (d.c.f.)	35	28	20	10	4

In upward c.f. the frequencies are cumulated starting from lowest class interval.

In downward c.f. the frequencies are cumulated starting from highest class interval.

For ungrouped Frequency Distribution, cumulative frequency (c.f.) of an observation is defined as the number of observations which are either equal to or smaller than that observation (upward c.f.) or the number of observations which are either equal to or greater than that observation (downward c.f.).

In the ungrouped Frequency Distribution given below, the upward & downward c.f. are obtained in the following manner:

Value of the variable (x_i)	5	7	9	14	18	24	27
Frequency (f_i)	3	6	4	10	2	4	7
Upward cumulative frequency (u.c.f.)	3	9	13	23	25	29	36
Downward cumulative frequency (d.c.f.)	36	33	27	23	13	11	7

For finding c.f. of grouped frequency distribution, the classes must be of exclusive type.

Rules for constructing frequency distribution

1. Classes must be clearly defined, exhaustive and mutually exclusive and preferably of equal width.
2. No. of classes should neither be too large nor too small. It should be usually between 5 and 15. It may go beyond 15 but can never be below 5.

   Number of class intervals is obtained by the

   (i) Sturge's Formula : $k = 1{+}3.322 \log_{10}N$

   (ii) Yule's Formula : $k = (2.5) \times N^{1/4}$;

   Where K is the number of class intervals

   N = Total Frequency

Class size = Range/k; Range means difference between the highest and lowest value of the data series.

Open-end class interval: In this type of class intervals, either the lower limit of first class interval or upper limit of last class interval or both are absent. Such types of classes are used when extreme values occur in data series. Such classes create difficulty in finding A.M., G.M., H.M., M.AD and Standard deviation. So median and mode can only be used as measure of average in open end classes but only Quartile deviation can be used as a measure of dispersion for these classes.

Income (In Rs)	< 18000	18000–21000	21000–24000	24000–27000	27000–30000	≥ 30000
No. of persons (frequency)	1	17	20	40	20	2

Extreme values are the values which differs widely from most of the values of the date series for e.g., If income of 97 persons (out of 100 persons) is within

Rs.18,000 to Rs.30,000 and for remaining 3 persons the incomes are Rs. 6,000, Rs.1,20,000 and Rs. 1,40,000 then these 3 values (Rs. 6,000, Rs.1,20,000 & Rs.1,40,000) are extreme values.

These data have to be represented by open-ended class intervals, otherwise many intermediate classes will have no frequencies, which is undesirable.

1. The data below shows the no. of fruits per tree for 60 trees 15, 16, 33, 45, 21, 25, 43, 22, 34, 21, 16, 21, 12, 25, 35, 40, 30, 33, 40, 27, 53, 41, 40, 35, 25, 22, 16, 25,15, 57, 25, 45, 57, 60, 65, 64, 38, 58, 61, 52, 39, 50, 60, 16, 62, 53, 57, 57, 43, 21, 14, 44, 45, 48, 62, 54, 57, 65, 44, 38. Prepare a suitable frequency distribution for the data.

Ans: The data are of discrete type. So the inclusive frequency distribution can be prepared for the data in the following manner.

First of all the no. of classes has to be determined by using sturge's formula or yule's formula. According to Sturge's Formula, no. of classes,

$$
\begin{aligned}
k &= 1 + 3.322 \log_{10}\text{N} \\
&= 1 + 3.322 \log_{10}60 \\
&= 1 + 3.322 \times 1.778 \\
&= 1 + 5.906 \\
&= 6.906 \approx 7
\end{aligned}
$$

Class size = range/k = (65 – 12)/7 = 7.57. Thus class size = 8

The inclusive frequency distribution is as follows:

Class Interval	10–17	18–25	26–33	34–41	42–49	50–57	58–65
Tally marks	𝍸 ||||	𝍸 𝍸 |	||||	𝍸 𝍸	𝍸 |||	𝍸 ||||	𝍸 ||||
Frequency	9	11	4	10	8	9	9

2. The data below shows the yield of mangoes per tree (in kg) for 45 mango trees 21.2, 44.7, 55.5, 43.2, 21.0, 20.7, 24.9, 25.3, 30.8, 32.2, 43.0, 55.1, 66.3, 54.2, 47.2, 65.1, 34.0, 21.1, 34.2, 55.3, 42.4, 21.1, 70.1, 69.5, 44.2, 56.0, 33.6, 44.3, 55.0, 66.2, 71.1, 22.7, 24.1, 34.9, 45.2, 37.3, 47.0, 67.4, 57.7, 55.1, 66.6, 77.8, 55.0, 55.5, 54.3. Prepare a suitable frequency distribution for the data.

   Ans: The data are of continuous type. So the exclusive frequency distribution can be prepared for the data in the following manner.

   First of all the no. of classes has to be determined by using sturge's formula or yule's formula. According to Sturge's Formula, no. of classes,

$$
\begin{aligned}
k &= 1 + 3.322 \log_{10}N \\
&= 1 + 3.322 \log_{10}45
\end{aligned}
$$

$$= 1 + 3.322 \times 1.653$$
$$= 1 + 5.491$$
$$= 6.491 \approx 7$$

Class size = range/k = (77.8–20.7)/7 = 8.16 Thus class size = 9

The inclusive frequency distribution is as follows:

Class Interval	19–28	28–37	37–46	46–55	55–64	64–73	73–82
Tally marks	𝍸 ||||	𝍸 |	𝍸 |||	𝍸 |	𝍸 ||	𝍸 |||	|
Frequency	9	6	8	6	7	8	1

Chapter 3

Presentation of Data

Presentation of data is done in form of diagram and graphs.

Diagrammatic Presentation: The data are presented in form of diagrams

The important diagrammatic presentations are simple bar diagram, multiple bar diagram, component bar diagram, percentage bar diagram & pie diagram.

Selection of diagram depends upon the nature of classification, number of characters involved and the type of comparison required to be made. In all the diagrams, the groups or classes are represented on the *X*-axis and the volumes or frequencies are represented in the *Y*-axis. The title of these visual aids should be self-explanatory.

1. **Simple Bar Diagram:** Simple bar diagrams consist of vertical bars of equal width. The height of these bars is proportional to the magnitude of the attribute. All bars are separated from each other by equal intervals. Simple bar diagram is used if the classification is based on attributes and if the attributes are to be compared with respect to a single character. For example, the area under different crops in a state, the production of mangoes in different years for a state, the yield of a variety at different fertilizer level, etc, can be represented by the simple bar diagram. Also the yield performance of different varieties of a crop, the effects of different treatments etc, can be compared using simple bar diagrams.

   A simple bar diagram for the production of rice crop in a farm for four different years: (Shown in Fig. 1)

Year	2014	2015	2016	2017
Production of rice crop (in Metric Tonnes)	36	44	60	50

2. **Component or Sub-Divided Bar Diagram:** Here the bars are sub-divided into different components. Each subdivision of the bars represents the component value of the character studied. This is used when each individual data can be sub-divided into its own components. Here each component of individual data are compared and also the individual data for each group can be compared. For example, the total production of a farm for three different crops grown in the farm in 4 years can be subdivided into its components. Here year is the attribute. The value (i.e., the total production) of a year is subdivided into three different components

i.e., production from three different crops in that year. Here the production from three different crops can be compared within a year and also the total production in different years can also be compared. Each component of an attribute are represented by different colours of different shadings but same colour or shading is used for a particular component in each group or attribute.

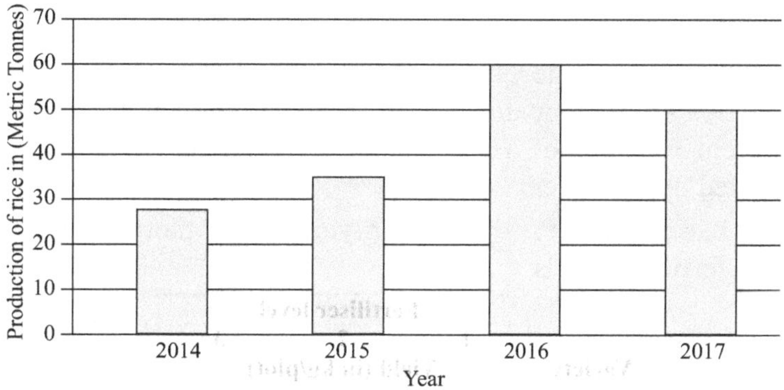

Fig. 1 Simple Bar showing production (in Metric Tonnes) ofrice in a farm for four years (2014–2017)

Component bar diagram for the production of three different crops in a farm in four years (2014–2017).

Year		2014	2015	2016	2017
Production (in Metric Tonnes)	Crop A	36	44	60	50
	Crop B	54	88	75	30
	Crop C	30	68	45	70
	Total	120	200	180	150

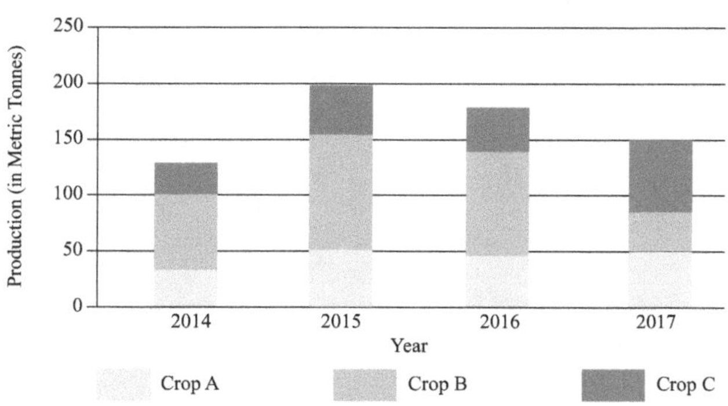

Fig. 2 Sub-divided Bar Diagram showing production (in Metric Tonnes) of three different crops grown in a farm four years (2014–2017)

3. **Multiple Bar Diagram:** If the groups or attributed are divided into two or more categories and the values of each category of a group are represented by different bars which are placed side by side without any gap and space is left between the bars of different groups (attributes), such type of diagrams are called multiple bar diagrams. Here multiple bars, placed side by side without any gap are present within a group. But in this case we cannot compare the total value of all the groups (obtained by combining the value of the categories in a group). Each bar within the group (attribute) will be marked or coloured differently in order to distinguish them. Same type of marking or colouring should be done for a particular category under each attribute. For e.g., the average yield of different varieties of a crop under different fertilizer level may be represented by multiple bar diagram.

   Multiple bar diagram for the data on yield (in kg/plot) of varieties for 3 different fertiliser levels:

	Fertiliser level		
	1	**2**	**3**
Variety	**Yield (in kg/plot)**		
A	25	28	33
B	34	44	46
C	65	68	60
D	44	48	42

   Multiple bar diagram is preferred when we want to compare only different categories within group (attribute) and not the total value of all categories of a group i.e., here groups are not compared.

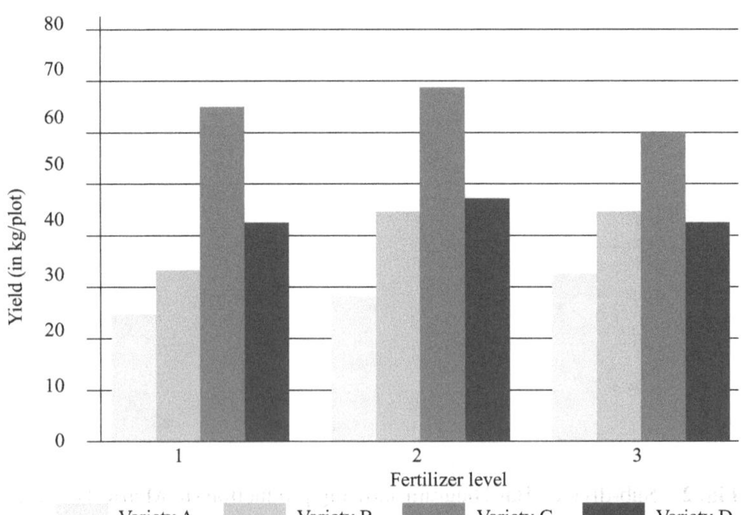

Fig. 3 Multiple Bar Diagrm showing yield of four different varieties for three different levels of fertilizer doses.

4. Percentage Bar Diagram: When the volumes of different attributes are greatly different for making meaningful comparisons the components of the attributes are reduced to percentages. Here each attribute will have 100 as its maximum volume. This type of component bar diagram is called percentage bar diagram.

Percentage bar diagram for the data on the production (in %) of crops *A*, *B*, *C* for four years (2014–2017).

Year		2014	2015	2016	2017
Production (in Metric Tonnes)	Crop A	36	42	63	51
	Crop B	54	90	72	30
	Crop C	30	68	45	69
	Total	120	200	180	150
Production (in %)	Crop A	30	21	35	34
	Crop B	45	45	40	20
	Crop C	25	34	25	46
	Total	100	100	100	100

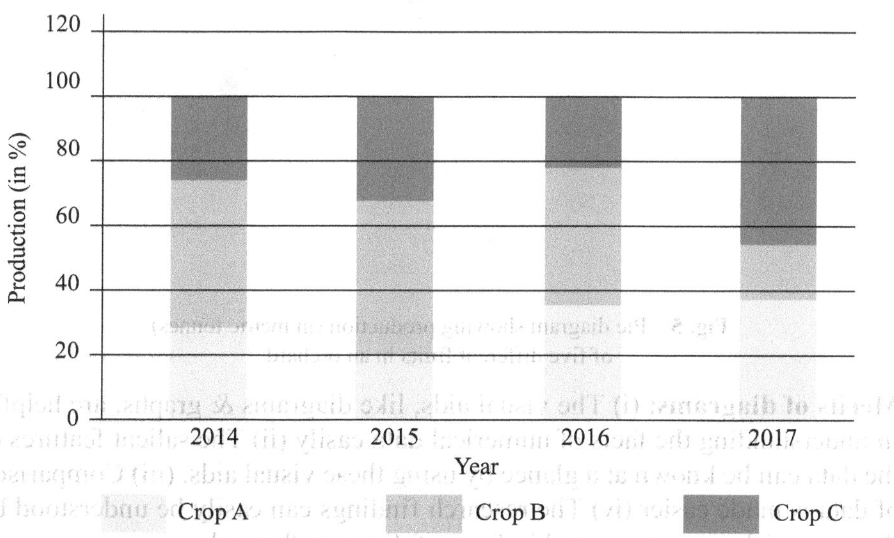

Fig. 4 **Percentage Bar diagram showing production(in %) of three different crops in a farm in four years (2014–2017)**

5. Pie Diagram– It is a circular diagram. Here the total value is represented in form of a circle (whose central angle is 360°) and the components are represented as the sectors with appropriate central angles, which is obtained as

$$\text{Central angle for a component} = \frac{\text{Value of a component}}{\text{Total value of all components}} \times 360^{\circ}$$

Example: Let the total production of fruits grown in a farm be represented by circle. The circle is partitioned into segments representing production of each fruit Area of the segments depend upon the value i.e., production of each fruit.

Pie diagram for the production (in Metric Tonnes) of different fruits grown in an orchard:

Fruits	**Mango**	**Coconut**	**Guava**	**Pine-apple**	**Custard-apple**
Production (in MT)	360	240	210	108	162
Central value of the component (in degrees)	120^0	80^0	70^0	36^0	54^0

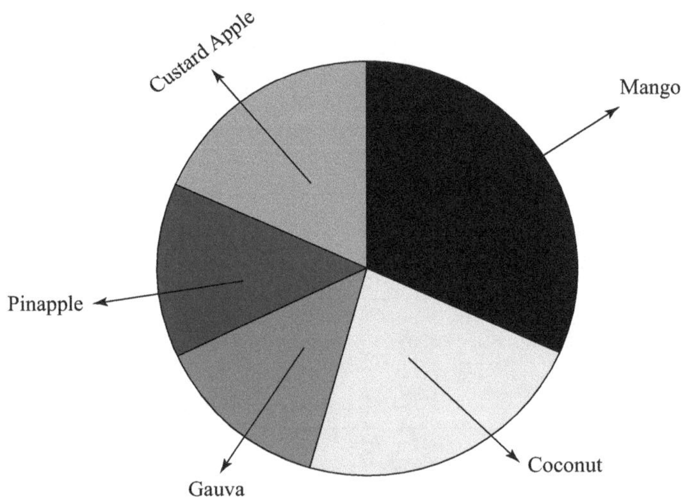

Fig. 5 Pie diagram showing production (in metric tonnes) of five different fruits in an orchard

Merits of diagrams: (i) The visual aids, like diagrams & graphs, are helpful in understanding the facts of numerical data easily (ii) The salient features of the data can be known at a glance by using these visual aids. (iii) Comparison of data is made easier (iv) The research findings can easily be understood by the users if they are presented in form of diagram & graphs.

Demerits: (i) Diagrams usually give approximate values. (ii) Becomes complicated when the no. of factors increases. If we have to represent a data involving the production of different varieties of different crops in different farms of region for some years, then the construction of diagram will become very difficult and the diagram will be very complicated.

Graphical Representation: The data are presented in form of graphs.

The various types of graphical presentation are histogram, frequency polygon, frequency curve and ogives. Selection of diagram depends upon the objective of study. In all the graphs, the groups or classes are represented on the X-axis and the volumes or frequencies are represented in the Y-axis. The title of the graphs should be self- explanatory.

A brief description of various types of graphical presentation is given below.

(i) **Histogram-** Histograms are the rectangles erected on a graph paper whose width corresponds to size of class interval and length/height corresponds to the class frequency (in case of equal class intervals) or corresponds to frequency density* (in case of unequal class intervals). There is no gap between the bars since the classes are continuous i.e., exclusive .Hence for constructing histogram classes must be continuous/ exclusive.

$$\text{* frequency density} = \frac{\text{frequency}}{\text{cs}}, \text{ where cs – class size}$$

In case of unequal class intervals, the area of the rectangle represents the frequency.

(ii) Frequency Polygon– It is also done for the data based on class-intervals, as like histogram. The frequencies of the classes are plotted by dots against the mid points of each class. The adjacent dots are then joined by straight lines, to obtain the frequency polygon.

Frequency polygon is used to compare two or more frequency distributions. Histogram cannot be used for such purposes. Also drawing of frequency polygon is simpler & quicker than drawing histograms.

***Line graphs:** These are frequency polygon type of graphs used to represent data which are not based on class intervals. Example, The food grain production over a period of time, growth rate of insects over a period of time, etc.

(iii) Frequency curve: If the class intervals, become smaller and smaller, resulting in increase of no. of class intervals, the frequency polygon tends to a smooth curve, called frequency curve. Area under the curve represents the total frequency.

It is easy to compare two frequency distributions through their frequency curves.

The frequencies of the classes are plotted by dots against the mid points of each class. The adjacent dots are then joined by free-hand smooth line to obtain the frequency curve.

(iv) **Ogive**: It is the graphical representation of cumulative frequency distribution, which may be plotted out of the upward c.f. distribution (for less than ogive.) or downward c.f. distribution (for more than ogive).

How to construct ogives

More than ogive: Points are plotted whose x- coordinate corresponds to lower class limit and y-coordinate corresponds to downward cumulative frequency. The points are joined by smooth lines to obtain more than ogive.

Less than ogive: points are plotted whose x-coordinate corresponds to upper class limit and y-coordinate corresponds to upward cumulative frequency. The points are joined by smooth lines to obtain less than ogive.

Construct a histogram for the following frequency distribution of yield/plot (in kg) of rice crop

Yield (in kg/plot) (Class Interval)	35–45	45–55	55–65	65–75	75–85	85–95	95–105
No. of plots (frequency)	25	35	40	50	45	30	20

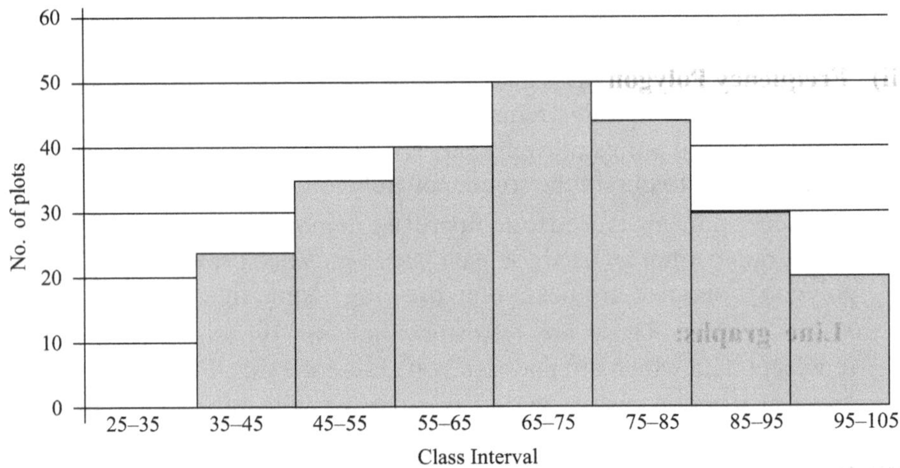

Fig. 6 Histogram

Construct a frequency polygon for the following frequency distribution of yield/plot (in kg) of rice crop

Yield (in kg/plot) (Class Interval)	35–45	45–55	55–65	65–75	75–85	85–95	95–105
No. of plots (frequency)	25	35	40	50	45	30	20

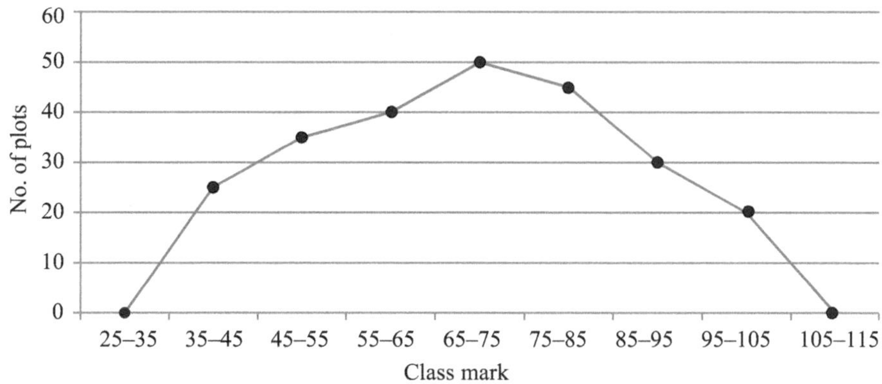

Fig. 7 Frequency Polygon

Construct a frequency curve for the following frequency distribution of yield/plot (in kg) of rice crop

Yield (in kg/plot) (Class Interval)	35–45	45–55	55–65	65–75	75–85	85–95	95–105
No. of plots (frequency)	25	35	40	50	45	30	20

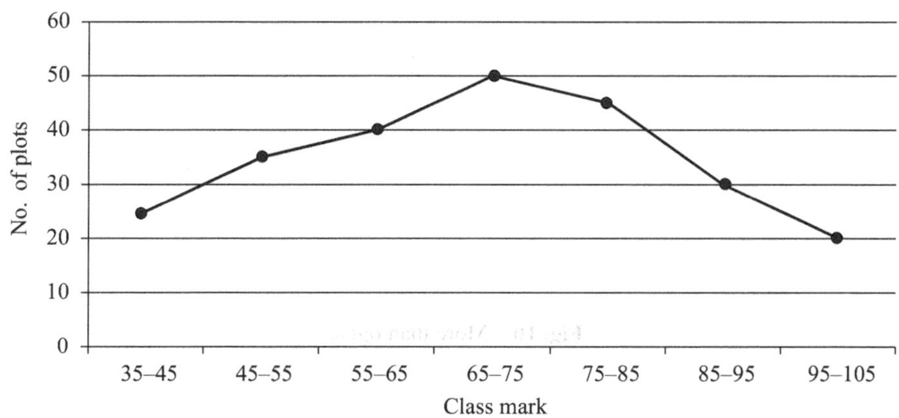

Fig. 8 Frequency Curve

For less than ogive, less than series has to be constructed:

Yield Less than	45	55	65	75	85	95	105
Number of plots (frequency)	25	60	100	150	195	225	245

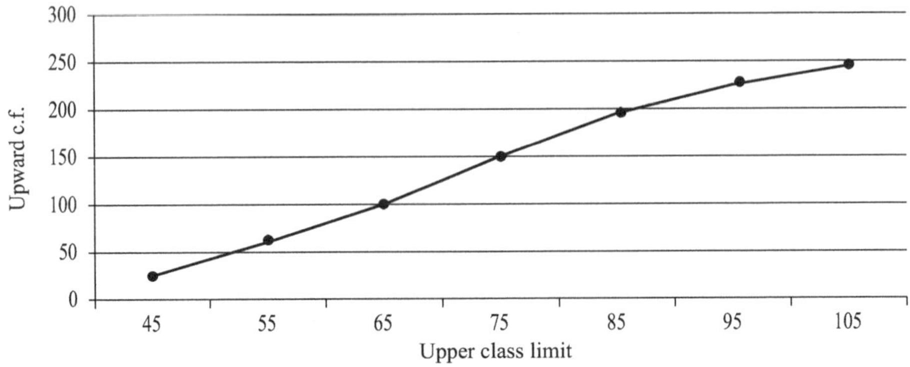

Fig. 9 Less than ogive

For more than ogive, less than series has to be constructed:

Yield More than or equal to	35	45	55	65	75	85	95
Number of plots (frequency)	245	220	185	145	95	50	20

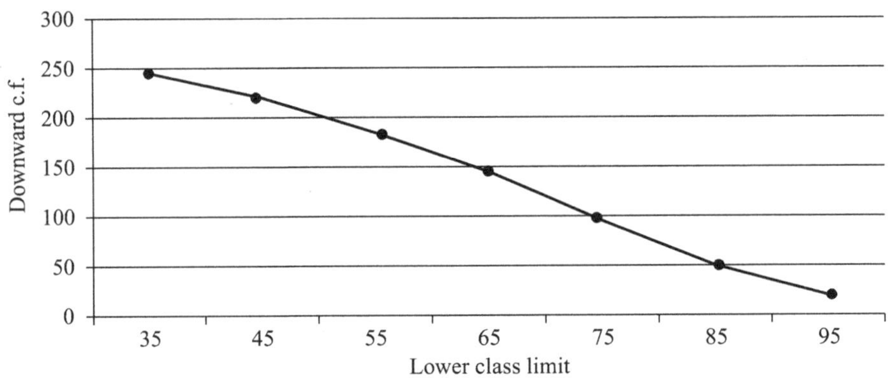

Fig. 10 More than ogive

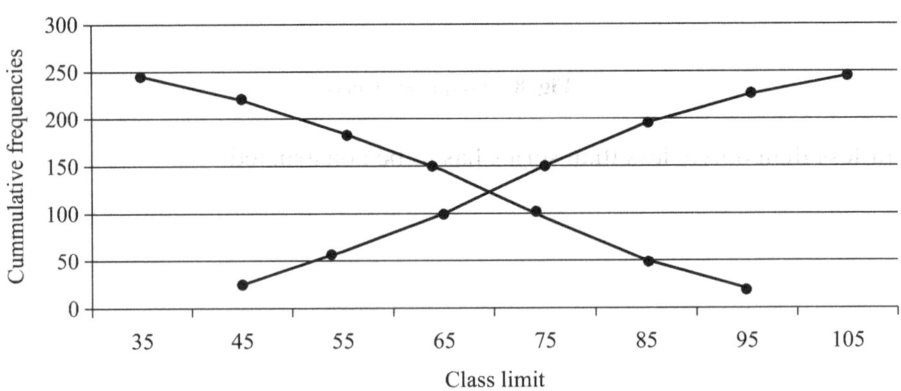

Fig. 11 Both ogives

Use of histogram to find mode

Mode can be obtained from histogram. A line is drawn diagonally joining top left corner of the highest rectangle to the top left of the succeeding rectangle. Another line is drawn diagonally joining the top right corner of the highest rectangle to the top right corner of the preceding rectangle. From the point of intersection of these two diagonals a perpendicular is drawn to x-axis. The point where this line meets x-axis gives the mode (Fig.12c).

Use of Ogive to find median and other partition values

The intersection of more than ogive and less than ogive gives the median (x-coordinate of the point of intersection gives the median). In case of less than ogive or more than ogive median is found by drawing a line perpendicular to y-axis at $N/2$ and from the point of intersection of the perpendicular with the ogive, a line is drawn perpendicular to x-axis. The point where this line meets x-axis gives the median. (Fig. 12a)

To find quartiles from less than ogive → A line perpendicular to y-axis is drawn from ($i \times N/4$) for i^{th} quartile and from the point of intersection of this line with ogive, the perpendicular is drawn to x-axis. The point where this perpendicular meets x-axis gives the ith quartile (Fig 12b).

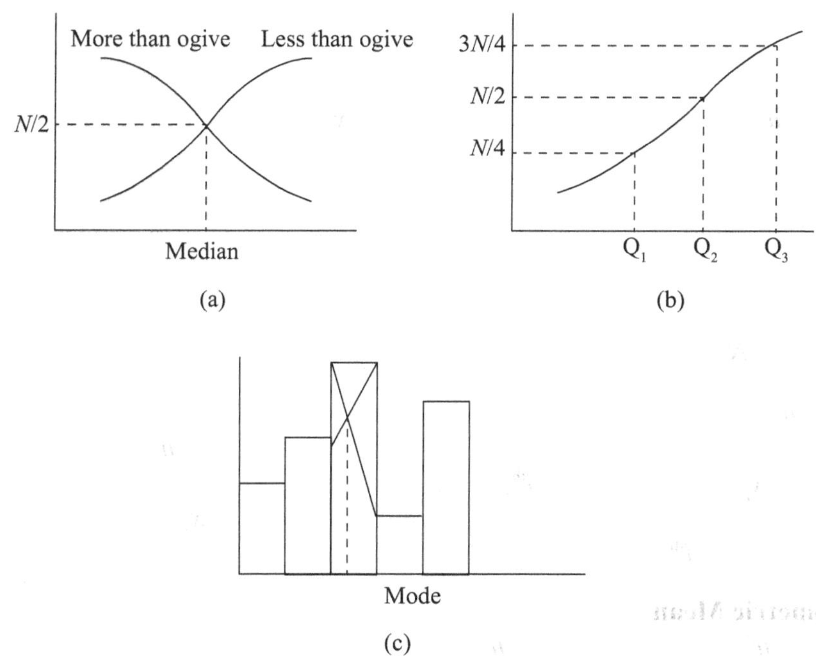

Fig. 12 (a) Use of both the ogives (a) to find median (b) less than ogive to find quartiles and (c) histograms to find mode

Chapter 4

Measures of Central Tendency

Measures of central tendency are also known as averages. Averages are defined as statistical constants which enable us to understand in a single effort, the significance of the whole i.e., they are the representative value of the distribution. They give an idea about the concentration of the values in the central part of the distribution.

Five Types of Averages

Arithmetic Mean

It is the sum of all the observations divided by the number of observation.

For individual data or simple series (i.e., series in which all values have single frequency),

$$\text{A.M.}(\bar{x}) = \frac{\sum_{i=1}^{n} x_i}{n}$$

Where n is the number of observations and x_i is the value of the observation.

For ungrouped and grouped frequency distribution

$$\text{A.M.}(\bar{x}) = \frac{\sum_{i=1}^{n} f_i x_i}{N}$$

Where N = Total frequency $= \sum_{i=1}^{n} f_i$

Here, n is the number of different observations in case of ungrouped frequency distribution. In case of grouped frequency distribution, n is the number of classes. x_i is the value of ith observation in case of ungrouped frequency distribution. In case of grouped frequency distribution x_i is the mid-value (or class mark) or i^{th} interval.

Geometric Mean

It is the nth root of product of "n" observation for individual data,

$$\bar{G} = (x_1 \times x_2 \times \cdots \times x_n)^{1/n} = \prod x_i^{1/n}$$

$$(\log \bar{G} = \log(\prod x_i^{1/n})$$

$$\bar{G} = \text{anti}\log\left(\frac{1}{n}\sum_{i=1}^{n}\log x_i\right)$$

For grouped and ungrouped frequency distribution:

$$\bar{G} = \left(x_1^{f_1} \times x_2^{f_2} \times \cdots \times x_n^{fn}\right)^{1/N}$$

$$\log \bar{G} = \frac{1}{N}\sum_{i=1}^{n} f_i \log x_i$$

$$\bar{G} = \text{anti}\log\left(\frac{1}{N}\sum_{i=1}^{n} f_i \log x_i\right)$$

Harmonic Mean

It is the reciprocal of the A.M. of the reciprocal of the observations.

For individual data;

$$\bar{H} = \frac{n}{\sum_{i=1}^{n}\frac{1}{x_i}}$$

For grouped and ungrouped frequency distribution;

$$\bar{H} = \frac{N}{\sum_{i=1}^{n}\frac{f_i}{x_i}}$$

For finding A.M., G.M., H.M., the inclusive cases may or may not be converted to exclusive.

Median: It is the value of the variables which exceeds and is exceeded by same number of observations i.e., it divides the entire data series into two equal parts. Thus median is a positional average. It is an ordinal measure of the central location of the data values.

How to find Median: For individual data series – The given data are arranged either in ascending or descending order of their magnitude. If 'n' is odd, then median = value of $\{(n + 1)/2\}^{th}$ observation.

If 'n' is even, then

median = [value of $(n/2)^{th}$ observation + value of $\{(n + 1)/2\}^{th}$ observation]/2

For ungrouped frequency distribution – median is the value whose cumulative frequency is either equal to or just greater than $N/2$ (if N is even) just greater than $(N + 1)/2$ (if N is odd).

For grouped frequency distribution – first we have to determine the median class. Median Class is the class whose cumulative frequency is either equal to or just greater than $N/2$.

$$\text{Median} = L + \left\{ \frac{\left(\frac{N}{2} - c \right)}{f} \right\} h$$

L = true lower limit of median class

f = frequency of median class

h = width of median class

c = cumulative frequency of the class just preceding the median class.

Partition Values

These are the values which divide the series into equal parts

They may be:

i. Quartiles – divides the series into 4 equal parts.
ii. Deciles – divides the series into 10 equal parts.
iii. Percentiles – divides the series into 100 equal parts.

Thus there are 3 quartiles, 9 deciles and 99 percentiles.

Mode – it is the value of the variable which has the highest frequency.

How to find Mode

For ungrouped Frequency Distribution, mode is the value having highest frequency.

For grouped frequency distribution:

$$\text{Mode} = L + \{h(f_1 - f_0)/(2f_1 - f_0 - f_2)\}$$

Modal class is the class having highest frequency.

L = true lower limit of the modal class;

f_1 = frequency of the modal class

f_0 = frequency of class just preceding the modal class

f_2 = frequency of class just succeeding the modal class

h = class size of the modal class.

If there are more than one mode in a data series, then a single approximate value of mode can be obtained from the empirical relationship:

$$\text{Mode} = 3 \times \text{Median} - 2 \times \text{Mean}.$$

For finding mode and median, the inclusive cases have to be converted to exclusive.

Properties of A.M.

1. Sum of the deviation of the observation from their mean is zero.

$$\sum_{i-1}^{n} (x_i - \overline{x}) = 0, \text{ for individual series}$$

$$\sum_{i-1}^{n} f_i(x_i - \overline{x}) = 0, \text{ for frequency Distribution.}$$

2. A.M. is affected both by change of origin and scale i.e., if $u_i = \frac{x_i - a}{h}$,
   then $\overline{x} = a + h\overline{u}$, where $\overline{u} = \frac{1}{N}\sum_{i=1}^{n} f_i u_i$
3. Capable for further statistical analysis.

If $x_1, x_2, \ldots, x_k$ are A.M. of K series of sizes $n_1, n_2, \ldots, n_k$ respectively, then their combined mean i.e., mean of combined series is given by,

$$\overline{x} = \frac{\sum_{i=1}^{k} n_i \overline{x}_i}{\sum_{i=1}^{k} n_i}$$

G.M. and H.M. are also suitable for further mathematical treatment:

If $\overline{G}_1, \overline{G}_2, \ldots, \overline{G}_k$ are G.M. of K- series of sizes $n_1, n_2, \ldots, n_k$ respectively then

their combined G.M. is given by $\overline{}$ $\text{Antilog}\left(\frac{\sum_{i=1}^{k} n_i \log \overline{G}_i}{\sum_{i-1}^{k} n_i}\right)$

If $\overline{H}_1, \overline{H}_2, \ldots, \overline{H}_k$ are H.M. of K-series of sizes $n_1, n_2, \ldots, n_k$ respectively then

their combined H.M. is given by $\overline{H} = \frac{\sum_{i=1}^{k} n_i}{\sum_{i=1}^{k} \frac{n_i}{\overline{H}_i}}$

Some other Properties of mean

$$\text{A.M.} \times \text{H.M.} = \text{G.M}^2.$$

A.M. $\geq$ G.M. $\geq$ H.M, the sign of equality holds good only if all values of the series are identical.

Characteristics of ideal average or ideal measure of central tendency

1. Should be easy to understand and calculate. The formula used for calculation of the measure should be very simple and easy for calculation.

2. Should be rigidly defined. This means for a particular set of observations there exist a unique value of the measure. For the data set 23, 43, 56 and 34 the mean cannot be any value other than 39.
3. Should be based on all observations. This means all the values of the data series should have been used for finding the measure.
4. Should be capable for further statistical analysis. Any measure of central tendency if suitable for further statistical analysis can give further information about the study for which the measure is used.
5. Should not be affected by extreme values. Extreme values are few values which differ widely from most of the values of the series. For example, let the yield of a crop in 147 plots out of 150 plots of a field from Rs. 200–300 kg . The yield of remaining three plots be 20 kg, 800 kg and 900 kg. These three values i.e., 20 kg, 800 kg and 900 kg differ widely from most of the values of the series. Thus, these three values are the extreme values of the series. An ideal measure of central tendency should not be affected by these extreme values.
6. Should be able to be located graphically. If a measure of central tendency can be located graphically then the value of the measure can be very quickly and easily known from the graph.

Merits of Arithmetic Mean

1. It is easy to understand and calculate.
2. It is rigidly defined.
3. It is based on all observations
4. It is capable for further statistical analysis

Demerits of Arithmetic Mean

1. It is affected very much by extreme values
2. It cannot be located graphically

Merits of Geometric Mean

1. It is rigidly defined
2. It is based on all observations
3. It is capable for further statistical analysis.

Demerits of Geometric Mean

1. It is not easy to understand and calculate
2. It cannot be located graphically
3. If any observation is zero, then geometric mean of the entire series becomes zero, which is absurd.

Merits of Harmonic Mean

1. It is rigidly defined
2. It is based on all observations
3. It is capable for further statistical analysis

Demerits of Harmonic Mean

1. It is not easy to understand and calculate
2. It cannot be located graphically
3. If any observation is zero, then harmonic mean cannot be found

Merits of Median

1. It is easy to understand and calculate
2. It is rigidly defined.
3. It is not affected by extreme values. This is because extreme values occur at the ends when the data are arranged in order of magnitude and median deals with observation at the middle position.
4. It is can be located graphically by the help of ogives.

Demerits of Median

1. It is not based on all observations. It is based on the observation at the middle position of the data series.
2. It is not capable for further statistical analysis. Median of two or more series cannot be used to find median of the whole series.

Merits of mode

1. It is easy to understand and calculate.
2. It is not affected by extreme values. This is because extreme values are only few in no. and mode deals with only the value(s) having maximum frequency.
3. It can be located graphically by the help of histogram.

Demerits of Mode

1. Mode may be ill-defined. In some cases there can exists more than one mode for a particular data series. A data series having one mode is called unimodal series, having two modes is called as bimodal series, having three modes is called trimodal series and having more than three modes is usually called as multimodal series.
2. It is not based on all observation. It is based on the observation which occurs for maximum no. of times in the data series.
3. It is not capable for further statistical analysis. Mode of two or more series cannot be used to find mode of the whole series.

Weighted Arithmetic Mean (Denoted as $\overline{w}$)

$$\overline{w} = \frac{\sum_{i=1}^{n} w_i x_i}{\sum_{i=1}^{n} w_i}$$; where w_i are the weights attached to the observation

If w_i's are equal, then $\overline{w} = \overline{x}$. If smaller values are given larger weights, then $\overline{w} < \overline{x}$.

If larger values are given larger weights, then $\overline{w} > \overline{x}$.

Use of weighted Arithmetic Mean

1. Used in calculating birth rate, death rates, average yield, index numbers etc.
2. It is used in such cases where the relative importance of the items varies and it is essential to allocate weights to the items.

Uses of Arithmetic Mean

1. It is the most widely used average due to its simplicity.
2. Used to find averages.
   For example, used to find average marks, average income, etc.

Uses of Geometric Mean

1. Used to find average growth rate.
2. Used in construction of index number.
3. Used to find the average relative change, as in case of bacterial growth, cell division etc.

Uses of Harmonic Mean

To find the average of the variables whose values are not obtained independently but derived from other variables, harmonic mean is to be used. For example, to find average yield whose value depends on values of area and production, harmonic mean is to be used. Also to find average speed whose value depends on values of distance and time, harmonic mean is to be used.

Uses of Mode

1. Used to find the average of qualitative data which has no numerical value and also cannot be ranked. For example, to find average cropping pattern of a region, the cropping pattern which is followed by most of the farmers (i.e., the modal value) will be the average one. To find average colour of rose flower in a garden where several rose flowers are grown, the colour which is found in maximum no. of flowers (i.e., the modal value) can be taken as the average colour.
2. Used in categorical data. To find average category of farmers from the categories like marginal, small, medium and large, the category to which

maximum farmers belong (i.e., the modal value) will be the average category.

Uses of Median

1. It is used when the extreme values of the series are not available or impossible to obtain.
2. It is essentially used in qualitative data where the observation is not susceptible to measurements in definite units but the data can be arranged in order on some basis.
3. Also useful in open-ended classes since median deals with the observations at middle position
4. It is very useful if there are wide variations in some observation of a series. In this case mean provides absurd results.
5. When rank is important, then median is used as average.

Numerical problems

1. Mean yield of rice in 30 plots of a farmer is 90 kg, in 70 plots of another farmer is 120 kg. Find the combined mean.

   Solution: Here, $\bar{x}_1 = 90$ kg; $\bar{x}_2 = 120$ kg; $n_1 = 30$; $n_2 = 70$

   Combined mean,

   $$\bar{x} = \frac{n_1\bar{x}_1 + n_2\bar{x}_2}{n_1 + n_2} = \frac{30 \times 90 + 70 \times 120}{30 + 70} = \frac{2700 + 8400}{100} = \frac{11000}{100} = 111$$

2. If mean and median of a series are 20 and 25 respectively, then find the mode of the series.

   Solution: Mode = 3 × Median – 2 × Mode = 3 × 25 – 2 × 20

   = 75 – 40 = 35

3. If A.M and G.M. of a data series are 50 and 30 respectively, then find H.M. of the series.

   Solution: A.M. × H.M. = G.M.^{2}

   $$50 \times \text{H.M.} = 30 \times 30 \Rightarrow \text{H.M.} = \frac{30 \times 30}{50} = 18$$

4. Mean of 7 observations is 66. If mean of first five observations is 60 and mean of last 3 observations is 60, then find the value of 5th observation.

   Solution: Total of 7 obs. = 7 × 66 = 462;

   Total of first five obs. = 5 × 60 = 300;

   Total of last 3 obs. = 3 × 60 = 180;

   Thus, value of 5th observation = 300 + 180 – 462 = 480 – 462 = 18

5. Find mean, G.M., H.M. and median of the following data: 12, 21, 23, 44, 45. Hence find the mode.

Solution:

x_i	12	21	23	44	45	$\sum_{i=1}^{n} x_i = 145$
log (x_i)	1.0792	1.3222	1.3617	1.6435	1.6532	$\sum_{i=1}^{n} \log(x_i) = 7.0598$
$\frac{1}{x_i}$	0.0833	0.0476	0.0435	0.0227	0.0222	$\sum_{i=1}^{n} \frac{1}{x_i} = 0.2193$

Here, no. of observations, $n = 5$

$$\text{Mean, } \bar{x} = \frac{\sum_{i=1}^{n} x_i}{n} = \frac{145}{5} = 29; \text{ G.M.,}$$

$$\bar{G} = \text{Antilog}\left(\frac{\sum_{i=1}^{n} \log(x_i)}{n}\right) \text{Antilog}\left(\frac{7.0598}{5}\right) = 25.82$$

$$\text{H.M., } \bar{H} = \frac{n}{1/x_i} = \frac{5}{0.2193} = 22.80$$

To find the median, first the data are to be arranged in order of magnitude. The data arranged in increasing order of magnitude in the following manner: 12, 21, 23, 44, 45.

Since, $n = 5$ (which is an odd number)

So, Median = value of (5 + 1)/2 th obs. = value of 3rd obs. = 23

Mode = 3 × median – 2 x Mean = 3 × 23 – 2 × 29 = 69 – 58 = 11

6. Find the A.M., G.M., H.M., Median and Mode of the given data:

x_i	30	35	45	50	60	70
f_i	8	12	20	24	16	10

Solution:

x_i	30	35	45	50	60	70	
f_i	8	12	20	24	16	10	$\sum_{i=1}^{n} f_i = 90$
f_ix_i	240	420	900	1200	960	700	$\sum_{i=1}^{n} f_i x_i = 4420$
$f_i\log(x_i)$	11.817	18.5288	33.0643	40.7753	28.4504	18.451	$\sum_{i=1}^{n} f_i \log(x_i) = 151.087$
f_i/x_i	0.2667	0.3429	0.4444	0.48	0.2667	0.1429	$\sum_{i=1}^{n} \frac{f_i}{x_i} = 1.9436$
c.f.	8	20	40	64	80	90	

n-the number of different observations; Here, $n = 6$

Mean, $\bar{x} = \frac{\sum_{i-1}^{n} f_i x_i}{N} = \frac{4420}{90} = 49.11;$

$$\text{G.M.}, \bar{G} = \text{Antilog}\left(\frac{\sum_{i-1}^{N} f_i \log x_i}{N}\right) = \text{Antilog}\left(\frac{151.087}{90}\right) = 47.72$$

H.M., $\bar{H} = \frac{N}{\sum_{i=1}^{N} \frac{f_i}{x_i}} = \frac{90}{1.9436} = 46.31;$

$N/2 = 90/2 = 45$; Median = 50 (since the c.f. 64 corresponding to the value 50, is greater than and nearest to 45)

Mode = 50 (since the value 50 has the highest frequency which is 24);

7. Find the A.M., G.M., H.M., Median and Mode of the given data:

Class Interval	20–30	30–40	40–50	50–60	60–70	70–80
f_i	6	8	12	10	9	5

Solution

Class Interval	20–30	30–40	40–50	50–60	60–70	70–80	
Class Mark (x_i)	25	35	45	55	65	75	
f_i	6	8	12	10	9	5	$\sum_{i=1}^{n} f_i = 50$
$f_i x_i$	150	280	540	550	585	375	$\sum_{i=1}^{n} f_i x_i = 2480$
$f_i \log(x_i)$	8.3876	12.3525	19.8386	17.4036	16.3162	9.3753	$\sum_{i=1}^{n} f_i \log(x_i) = 83.6738$
f_i/x_i	0.24	0.2286	0.2667	0.1818	0.1385	0.0667	$\sum_{i=1}^{n} \frac{f_i}{x_i} = 1.1223$
c.f.	6	14	26	36	45	50	

Where n – number of class intervals; N -Total number of observations;

$n = 6; N = \sum_{i=1}^{n} f_i = 50.$

Measures of Central Tendency

Mean, $\bar{x} = \frac{\sum_{i-1}^{n} f_i x_i}{N} = \frac{2480}{50} = 49.60;$

G.M., $\overline{G} = \text{Antilog}\left(\frac{\sum_{i-1}^{n} f_i \log x_i}{N}\right) = \text{Antilog}\left(\frac{83.6738}{50}\right) = 47.15$

H.M., $\overline{H} = \frac{N}{\sum_{i=1}^{n} \frac{f_i}{x_i}} = \frac{50}{1.1223} = 44.55;$

Mode $= L + h \times \left(\frac{f_1 - f_0}{2f_1 - f_0 - f_2}\right)$, Where L-True Lower Limit of modal class; h-size of modal class; f_1-frequency of modal class; f_2-frequency of class succeeding the modal class; f_0–frequency of class preceding the modal class.

Modal class is the class having highest frequency. Here the class 40–50 is the modal class as it has the highest frequency which is 12. $L = 40$, $h = 50 - 40 = 10, f_0 = 8, f_1 = 12, f_2 = 10$

Mode $= 40 + 10 \times \left(\frac{12-8}{2x12-8-10}\right) - 40 + 10 \times \frac{4}{6} = 40 + 6.67 = 46.67$

Median $= L + h \times \left(\frac{\frac{N}{2} - C}{f}\right)$; Where L-True lower limit of median class; f-frequency of the median class; C-the c.f. less than and nearest to $N/2$.

Here, $\frac{N}{2} = \frac{50}{2} = 25$

Thus the median class is 40–50.

So, $L = 40$, $C = 14$, $f = 12$, $h = 50 - 40 = 10$

Thus, median $= 40 + 10 \times \left(\frac{25-14}{12}\right) = 40 + 9.11 = 49.11$

8. Find the weighted A.M. for the data:

x_i	23	25	45	36	44
w_i	3	2	1	4	5

Solution:

x_i	23	25	45	36	44	
w_i	3	2	1	4	5	$\sum_{i=1}^{5} w_i = 15$
$w_i x_i$	69	50	45	144	220	$\sum_{i=1}^{5} w_i x_i = 528$

$$\text{Weighted A.M., } \overline{w} = \frac{\sum_{i=1}^{5} w_i x_i}{\sum_{i=1}^{5} w_i} = \frac{528}{15} = 35.2$$

9. If 20 is subtracted from each of the observations and then divided by 5, then how the new mean is affected?

   Solution: Let x_i be the original variable having values $x_1, x_2, \ldots, x_n$.

$$\text{Let } u_i = \frac{x_i - 20}{5}$$

$$\Rightarrow \sum_{i=1}^{n} u_i = \sum_{i=1}^{n} \frac{x_i - 20}{5}$$

$$\Rightarrow \sum_{i=1}^{n} u_i = \frac{\sum_{i=1}^{n} x_i - n \times 20}{5}$$

$$\Rightarrow \frac{\sum_{i=1}^{n} u_i}{n} = \frac{\sum_{i=1}^{n} x_i - n \times 20}{n \times 5}$$

$$\Rightarrow \overline{u} = \frac{\overline{x} - 20}{5}$$

   Thus the new mean is obtained by subtracting 20 from the original mean and then dividing by 5.

Chapter 5

Measures of Dispersion, Moments, Skewness and Kurtosis

Literal meaning of dispersion is scatterdness. We study dispersion to have an idea about the homogeneity and heterogeneity of the distribution. Dispersion gives an idea about how much the values of the observation varies from their averages. The series in which the values of the observation differ much from their averages is said to be heterogeneous or highly dispersed. The series in which the value of the observation does not differ much from their averages is said to be homogenous or less dispersed.

There are two types of measures of dispersion:

1. **Absolute measures of dispersion:** Range, mean absolute deviation, quartile deviation, standard deviation are the absolute measures of dispersion. The absolute measures of dispersion carry the same unit of the observations.
2. **Relative measures of dispersion:** These measures are the ratio of the measure of dispersion to the corresponding measure of central tendency. These measures have no units. They are also known as coefficient of Dispersion. The measures are used to compare the variability in two or more series mainly when they are expressed in different units. The various measure of relative measures of dispersion are :
   i. Coefficient of Dispersion (C.D.) based on range
   ii. C.D. based on Mean Absolute Deviation.
   iii. C.D. based on Quartile Deviation.
   iv. C.D. based on Standard Deviation or Coefficient of Variation

Coefficient of variation is the important relative measure of dispersion.

1. **Range:** It is the difference between largest and smallest observation of a series.

   Range = Largest – Smallest

   When slight variation can create great difference, then range is used .e.g., pH of the soil, temperature of soil.

2. **Mean Absolute Deviation (M.A.D):** It is the A.M. of the absolute deviation of the observations from their average (mean, median and mode).

   M.A.D $= \frac{1}{n}\sum_{i=1}^{n}|x_i - A|$; for individual series

   M.A.D $= \frac{1}{N}\sum_{i=1}^{n} f_i|x_i - A|$; for frequency distribution,

   Where A = mean, median or mode

3. **Quartile Deviation:**(Used for qualitative data)

   It is half of the difference between third and first quartile.

   $$Q.D. = \frac{1}{2}(Q_3 - Q_1)$$

   We use quartile deviation when variation in ranking is present.

   It is also known as semi-inter quartile range.

4. **Standard Deviation**

   It is defined as the positive square root of the A.M of the square of deviations of the observations from their mean. Denoted as σ.

   $\sigma = +\sqrt{\frac{1}{n}\sum_{i=1}^{n}(x_i - \overline{x})^2}$; for individual data.

   The shortcut formula is $\sigma = +\sqrt{\frac{1}{n}\sum x_i^2 - \left[\frac{\sum_{i=1}^{n} x_i}{n}\right]^2}$

   $\sigma = +\sqrt{\frac{1}{N}\sum_{i=1}^{n} f_i\left(x_i - \overline{x}\right)^2}$; for frequency distribution

   The shortcut formula is $\sigma = +\sqrt{\frac{1}{N}\sum_{i=1}^{n} f_i x_i^2 - \left[\frac{\sum_{i=1}^{n} f_i x_i}{N}\right]^2}$

   Square of S.D. is known as Variance. It is denoted by σ^2

   Variance is defined as A.M of the square of deviations of the observations from their mean.

Merits and Demerits of various measures of dispersion

Merits of Range: It is very easy to understand and calculate.

Demerits of Range

1. Since it is based on two extreme observations which themselves are subjected to chance fluctuations, it is not at all reliable measure of dispersion.

2. Not suitable for further statistical analysis

Merits of M.A.D

1. It is based on all observation.
2. Easy to understand and calculate

Demerits of M.A.D

1. Since the sign of the deviations are omitted by taking absolute value, it is mathematically incorrect i.e., it has abstract mathematical character.
2. Not suitable for further statistical analysis

Merits of Quartile Deviation

1. Easy to understand and calculate.
2. Not affected by extreme values.

Demerits of Quartile Deviation

1. It is not based on all observations i.e., it is based on only 50% of the observation as it omits first 25% and last 25% of the series
2. It is not suitable for further statistical analysis

Merits of Standard Deviation/Variance

1. It is suitable for further statistical analysis
2. It is based on all observations
3. It overcomes the limitations of M.A.D by taking squares of the deviations to ignore the sign of deviation.

Demerits of Standard Deviation/Variance

1. It is difficult to calculate
2. Affected much by extreme values

Ideal and most widely used measure of dispersion is standard deviation (σ) or variance (σ^2).

Properties of S.D (σ) / variance (σ^2)

1. It is independent of change of origin but not of scale.

   If $u_i = \dfrac{x_i - a}{h}$, then i.e., $\sigma_U^2 = \dfrac{\sigma_X^2}{h^2} \Rightarrow \sigma_U = \dfrac{\sigma_X}{h}$

2. It is suitable for further statistical analysis.

   If $\sigma_1, \sigma_2, \sigma_3, \cdots, \sigma_k$ are S.D. and $\bar{X}_1, \bar{X}_2, \bar{X}_3, \cdots, \bar{X}_k$ means of K-series of sizes n_1, n_2, n_3, ... n_k respectively then the S.D. of the combined series is given by:

$$= +\sqrt{\frac{\sum n_i(\sigma_i^2 \quad d_i^2)}{\Sigma}} \quad \text{where} = d_i = \overline{x}_i - \overline{x},\ \overline{x}_1, \overline{x}_2, \overline{x}_3 \ldots \overline{x}_k$$

Relative measures of dispersion: It is obtained by dividing absolute measure of dispersion with appropriate measure of average. Also known as coefficient of Dispersion.

$$\text{Relative measure of dispersion} = \frac{\text{absolute measure of dispersion}}{\text{measure of central tendency}}$$

Relative measure of dispersion is used when we are comparing the variability of two or more series. Absolute measure of dispersion is used to find variability in a series.

1. Coefficient of Dispersion (C.D.) based on range $= \dfrac{L-S}{L+S}$;
   L–Largest observation; S–Smallest observation
2. Coefficient of Dispersion (C.D.) based on M.A.D. $= \dfrac{\text{M.A.D.}}{\text{A}}$;

   Where, A = mean/median/mode

3. Coefficient of Dispersion (C.D.) based on

$$\text{Q.D.} = \frac{\frac{1}{2}(Q_3 - Q_2) - (Q_2 - Q_1)}{\frac{1}{2}(Q_3 - Q_2) + (Q_2 - Q_1)}$$

$$= \frac{(Q_3 - 2Q_2 + Q_1)}{(Q_3 - Q_1)}$$

3. Coefficient of Dispersion (C.D.) based on Standard Deviation or Coefficient of Variation (C.V) is the most important relative measure of dispersion.
4. Definition of C.V: It is defined as standard deviation as a percentage of mean.

$$\text{C.V} = \frac{\sigma}{\overline{x}} \times 100$$

Uses of C.V: Used to compare variability in two or more series.

Interpretation of C.V: Series having high C.V. is said to be more heterogeneous/ less consistent/more inconsistent. Series having low C.V. is said to be more homogeneous/more consistent/less inconsistent.

Covariance –When two variables are studied simultaneously, the variation between the two variables is known as covariance. Covariance measures the extent of variation between the observations of two variables. Covariance between two variables X and Y is given as,

$$\text{Cov}(x, y) \text{ or } \sigma_{XY} = \frac{1}{n}\sum_{i=1}^{n}(x_i - \bar{x})(y_i - \bar{y});\ \text{for simple series.}$$

$$\text{Cov}(x, y) \text{ or } \sigma_{XY} = \frac{1}{N}\sum_{i=1}^{n} f_i(x_i - \bar{x})(y_i - \bar{y});\ \text{for frequency distribution.}$$

Covariance is independent of origin but not of scale.

If $u_i = \frac{x_i - a}{h}$ and $v_i = \frac{y_i - b}{k}$ then $\sigma_{uv} = \frac{\sigma_{xy}}{h \times k}$

Thus covariance is the variation between two series of observations or two variables, whereas, variance is the variation within a single series or single variable.

Difference between C.V. (Coefficient of Variation) and Covariance

C.V. is the relative measure of variation within a single series or single variable, whereas covariance is the absolute measure of variation between two series or two variables. C.V. is unitless. It is expresses as %. Covariance takes the units of the variables.

Moments

The rth moment of a variable X about any point A is given by:

$$'_r = -\Sigma (x_i - A)\ ;\ \text{for individual or simple series}$$

$$\mu'_r = \frac{1}{N}\sum_{i=1}^{n} f_i(x_i - A)^r;\ \text{for frequency distribution}$$

where 'A' is any number. 'r' is the order of the moment.

If $A = \bar{x}$, then it is known as r^{th} central moment and is represented as m_r.

If A is any number other than $\bar{x}$ (mean) then it is known as r^{th} raw moment and is denoted as μ'_r.

If $A = 0$, it is called r^{th} raw moment about origin.

$$\mu_0 = \frac{1}{N}\sum_{i=1}^{n} f_i(x_i - \bar{x})^0 = 1;\ \mu_1 = \frac{1}{N}\sum_{i=1}^{n} f_i(x_i - \bar{x})^1 = 0;$$

$$\mu_2 = \frac{1}{N}\sum_{i=1}^{n} f_i(x_i - \bar{x})^2 = \sigma^2$$

i.e., 2nd central moment is the variance.

$$\mu_3 = \frac{1}{N}\sum_{i=1}^{n} f_i(x_i - \overline{x})^3;\ \mu_4 = \frac{1}{N}\sum_{i=1}^{n} f_i(x_i - \overline{x})^4$$

First raw moment (about A), $\mu_1' = \frac{1}{N}\sum_{i=1}^{n} f_i(x_i - A)^1 = \frac{1}{N}\sum_{i=1}^{n} f_i x_i - \frac{1}{N}\sum_{i=1}^{n} f_i A$

Thus first raw moment can give the mean.

If $A = 0$, then $\mu_1' = \overline{X}$ i.e., first raw moment about origin gives the mean.

Skewness: Lliteral meaning of skewness is 'lack of symmetry.' Dispersion measures the scatteredness in data, whereas, skewness provides the direction of scatteredness. Symmetry means there is no skewness in data i.e., the value of observation deviates equally in both the direction. The number of observations hat are greater than mean are almost equal to the no. of observations that are smaller than mean and the medium values of the series have greater frequency. Here the frequency curve is bell shaped or symmetrical. Here, mean = median = mode. Example: yield of a crop in kg /plot where the crop management is of average category.

If the curve extends more towards right, it is called right tailed curve and here the skewness is positive. Here most of the value of the obs. are less than the mean and the low values of the series have greater frequency Here, mean > median > mode. Example- Income distribution of people of India, yield of a crop in kg /plot where the crop management is of poor category

If the curve extends more towards left, it is called left tailed curve & here the skewness is negative. Here most of the value are greater than mean and the high values of the series have greater frequency. Here, mean < median < mode. The distribution of age at death for people of India, yield of a crop in kg /plot where the crop management is of high category

Measure of skewness is provided by $\beta_1 \quad \mu_3^2\ \mu_2^3$ but provided more appropriately by $\gamma_1 = \mu_3/\mu_2^{3/2}\,\beta_1$ is called coefficient of skewness.

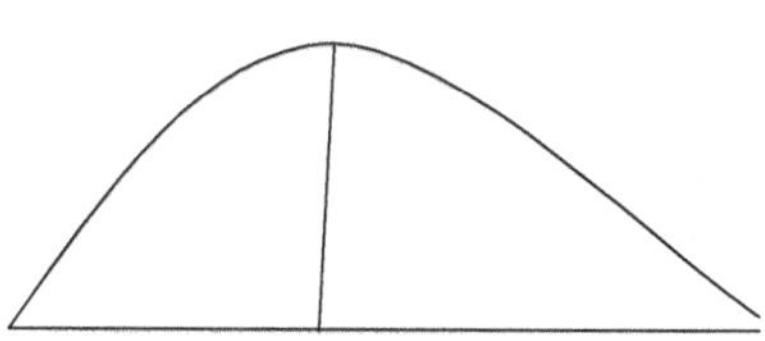

Mean = Median = Mode
Symmetrical Bell shape curve
(skewness = 0) $\beta_1 = 0, \gamma_1 = 0$

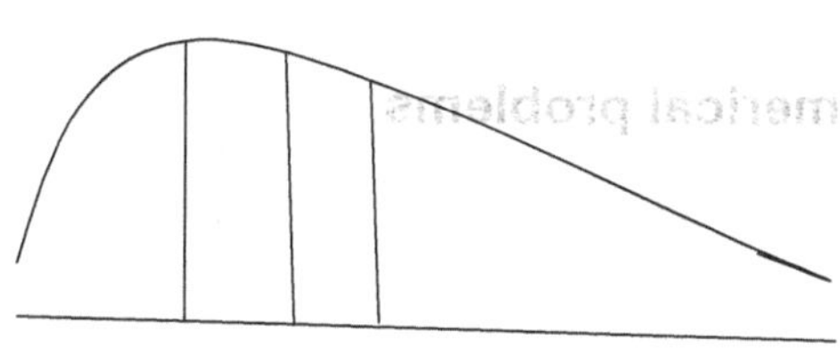

Mean >Median>Mode
Right tailed curve
(positive skewness), $\gamma_1 > 0$

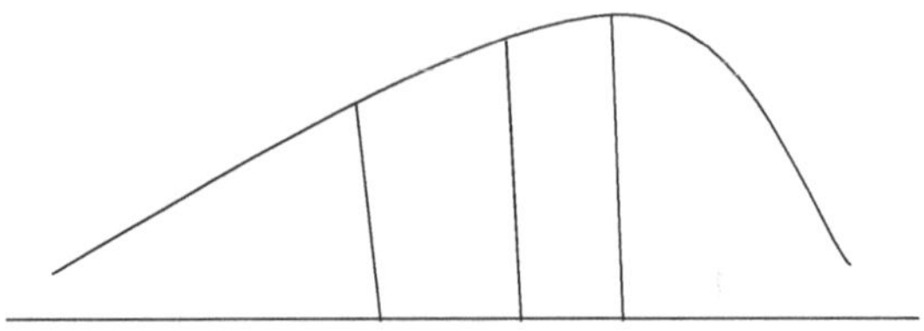

Kurtosis: It means peakedness. It determines peakedness of the frequency curve.β_2 and γ_2 provides measure of Kurtosis. $\beta_2 = \mu_4/\mu_2^2; \gamma_2 = \beta_2 - 3$

If the curve is more flatter, then it is called platykurticcurve. Here, $\gamma_2 < 0$; $\beta_2 < 3$.If the curve is more peaked, then it is called leptokurtic curve. Here, $\gamma_2 > 0$; $\beta_2 > 3$. If the curve is neither peaked nor flat it is called mesokurtic curve. Here $\gamma_2 = 0; \beta_2 = 3$

The curve which is both symmetric and mesokurtic is called normal curve.

Examples:

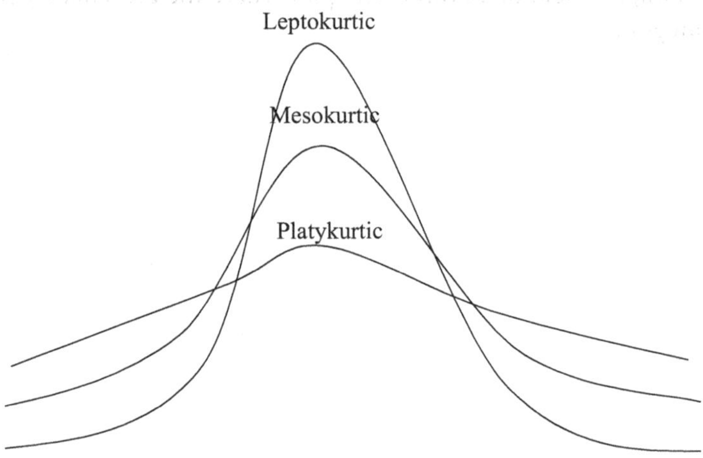

When a curve is symmetric, it may or may not be mesokurtic.

Numerical problems

1. For the following data find the combined mean and standard deviation. Also find out which series is more heterogeneous.

Series	Size	Mean	Standard Deviation
A	6	20	8
B	14	30	16

Solution: Combined mean, $\overline{x} = \dfrac{n_1\overline{x}_1 + n_2\overline{x}_2}{n_1 + n_2} = \dfrac{6 \times 20 + 14 \times 30}{6 + 14} = \dfrac{540}{20} = 27$

Combined standard deviation, $\sigma = \sqrt{\dfrac{n_1(\sigma_1^2 + d_1^2) + n_2(\sigma_2^2 + d_2^2)}{n_1 + n_2}}$

$$d_1 = \bar{x}_1 - \bar{x} = 20 - 27 = -7;\ d_2 = \bar{x}_2 - \bar{x} = 30 - 27 = 3$$

$$\therefore \sigma = \sqrt{\frac{6(64+49)+14(256+9)}{6+14}} = \sqrt{\frac{678+3710}{20}} = \sqrt{\frac{4388}{20}} = 14.81$$

$$CV_A = \frac{\sigma_A}{\bar{x}_A} \times 100 = \frac{8}{20} \times 100 = 40\%;$$

$$CV_B = \frac{\sigma_B}{\bar{x}_B} \times 100 = \frac{16}{30} \times 100 = 53.33\%$$

Since series *B* has larger *CV*, Series *B* is more heterogeneous.

2. If the first raw moment about 6 is 12, then find the mean.

Solution

$$\mu'_{1(\text{about}6)} = \text{mean} - 6 \Rightarrow 12 = \text{mean} - 6 \Rightarrow \text{mean} = 12 + 6 = 18$$

3. Calculate the measures of central tendency (Arithmetic Mean, Median, Mode, Quartiles and Deciles), measures of dispersion (*QD*, Standard Deviation, Variance and CV), measures of moments (μ_r; r = 1, 2, 3, 4), measures of skewness and measures of kurtosis for the following grouped data regarding yield of green gram (in kg) per plot for 150 plots.

Yield (in kg) (Class Interval)	30–35	35–40	40–45	45–50	50–55	55–60	60–65	65–70
No. of plots (frequency)	10	20	24	28	24	30	14	6

Solution:

Yield (in kg) (Class Interval)	**No. of plots (frequency, f_i)**	**Class mark (x_i)**	c.f.	$f_i.x_i$	$f_i . \log x_i$	f_i/x_i	$f_i . x_{i2}$	$f_i(x_i - \bar{x})^3$	$f_i(x_i - \bar{x})^4$
30–35	10	32.5	10	325	15.1188	0.3077	10562.5	–47372.91	795621.98
35–40	20	37.5	30	750	31.4806	0.5333	28125	–32817.82	387081.93
40–45	24	42.5	54	1020	39.0813	0.5647	43350	–7529.31	51160.68
45–50	28	47.5	82	1330	46.9474	0.5895	63175	–161.9	290.6
50–55	24	52.5	106	1260	41.2838	0.4571	66150	790.22	2532.75
55–60	30	57.5	136	1725	52.79	0.5217	99187.5	16572.09	135976.15
60–65	14	62.5	150	875	25.1423	0.224	54687.5	32237.1	425694.98
65–70	6	67.5	156	405	10.9758	0.0889	27337.5	36201.99	659061.92
Total	$\sum_{i=1}^{n} f_i = 156$			$\sum_{i=1}^{n} f_i x_i = 7690$	$\sum_{i=1}^{n} f_i \log x_i = 262.82$	$\sum_{i=1}^{n} f_i/x_i = 3.2869$	$\sum_{i=1}^{n} f_i x_i^2 = 392575$	$\sum_{i=1}^{n} f_i(x_i - \bar{x})^3 = 2080.54$	$\sum_{i=1}^{n} f_i(x_i - \bar{x})^4 = 2457521$

Where n – number of class intervals; N-Total number of observations;

$$n = 8;\ N = \sum_{i=1}^{n} f_i = 156$$

Measures of Central Tendency:

Mean, $\bar{x} = \dfrac{\sum_{i-1}^{n} f_i x_i}{N} = \dfrac{7690}{156} = 49.29$; G.M.,

$$\text{G.M., } \bar{G} = \text{Antilog}\left(\frac{\sum_{i=1}^{N} f_i \log x_i}{N}\right) = \text{Antilog}\left(\frac{262.82}{156}\right) = 48.39$$

$$\text{H.M., } \bar{H} = \frac{N}{\sum_{i=1}^{n} \frac{f_i}{x_i}} = \frac{156}{3.2869} = 47.46$$

Mode $= L + h \times \left(\dfrac{f_1 - f_0}{2f_1 - f_0 - f_2}\right)$, Where L – True Lower Limit of modal class; h – Size of modal class; f_1 – frequency of modal class; f_2 – frequency of class succeeding the modal class; f_0 – frequency of class preceding the modal class.

Modal class is the class having highest frequency. Here the class 55 – 60 is the modal class as it has the highest frequency which is 30. $L = 55$, $h = 60 - 55 = 5$, $f_0 = 24$, $f_1 = 30$, $f_2 = 14$

$$\text{Mode} = 55 + 5 \times \left(\frac{30 - 24}{2x30 - 24 - 14}\right) = 55 + 5 \times \frac{6}{22} = 55 + 1.36 = 56.36$$

Median $= L + h \times \left(\dfrac{\frac{N}{2} - C}{f}\right)$; Where L – True lower limit of median class; f– frequency of the median class; C – the c.f. less than and nearest to $N/2$.

Here $\dfrac{N}{2} = \dfrac{156}{2} = 78$; Thus the median class is 45–50.

So, $L = 45$, $C = 54$, $f = 28$; $h = 5$;

$$\text{Thus median } = 45 + 5 \times \left(\frac{78 - 54}{28}\right) = 45 + 4.29 = 49.29$$

Quartiles: i^{th} quartile, $Q_i = L + h \times \left(\dfrac{\frac{i \times N}{4} - C}{f}\right)$, Where, L-True lower limit of ith quartile class, $i = 1$ for 1st quartile (Q_1), $i = 2$ for 2nd quartile (Q_2), $i = 3$ for

3rd quartile (Q_3); i^{th} quartile class is the class whose c.f. is equal to or greater than and nearest to $i \times \frac{N}{4}$; f–frequency of the ith quartile class; C–the c.f. less than and nearest to $i \times \frac{N}{4}$

$$1 \times \frac{N}{4} = 1 \times \frac{156}{4} = 39; 2 \times \frac{N}{4} = 2 \times \frac{156}{4} = 78; 3 \times \frac{N}{4} = 3 \times \frac{156}{4} = 117$$

Class whose c.f. is either equal to or greater than and nearest to 39 is 40–45;
Class whose c.f. is either equal to or greater than and nearest to 78 is 45–50;
Class whose c.f. is either equal to or greater than and nearest to 117 is 55–60;

For 1st Quartile, 1st quartile class is 40–45; $L = 40$, $C = 30$, $f = 24$, $Q_1 = 40 + 5 \times \left(\frac{39-30}{20}\right) = 40 + 2.25 = 42.25$.

For 2nd Quartile, 2nd Quartile class is 45–50; $L = 45$, $C = 54$, $f = 28$, $Q_2 = 45 + 5 \times \left(\frac{78-54}{28}\right) = 45 + 4.29 = 49.29$.

For 3rd Quartile, 3rd Quartile class is 55–60; $L = 55$, $C = 106$, $f = 30$, $Q_3 = 55 + 5 \times \left(\frac{117-106}{30}\right) = 55 + 1.83 = 56.83$

****2nd Quartile is same as median**

Deciles: i^{th} Decile, $D_i = L + h \times \left(\frac{\frac{i \times N}{10} - C}{f}\right)$, Where L – True lower limit of ith Decile class, $i = 1$ for 1st Decile (D_1), $i = 2$ for 2nd Decile (D_2), …, $i = 5$ for 5th Decile (D_5), … $I = 9$ for 9th Decile (D_9)

i^{th} Decile class is the class whose c.f. is equal to or greater than and nearest to $i \times \frac{N}{10}$;

f – frequency of the ith Decile class; C – the c.f. less than and nearest to $i \times \frac{N}{10}$

$$1 \times \frac{N}{10} = 1 \times \frac{156}{10} = 15.6;$$

$$2 \times \frac{N}{10} = 2 \times \frac{156}{10} = 31.2, \cdots 5 \times \frac{156}{10} = 78, \cdots 9 \times \frac{N}{10}$$

$$= 9 \times \frac{156}{10} = 140.4$$

Class whose c.f. is either equal to or greater than and nearest to 15.6 is 35–40;
Class whose c.f. is either equal to or greater than and nearest to 31.2 is 40–45;

Class whose c.f. is either equal to or greater than and nearest to 78 is 45–50;
Class whose c.f. is either equal to or greater than and nearest to 140.4 is 60–65.

For 1st Decile, 1st Decile class is 35–40; $L = 35$, $C = 10$, $f = 20$; $D_1 = 35 + 5 \times \left(\frac{15.6 - 10}{20}\right) = 35 + 1.4 = 36.4$

For 2nd Decile, 2nd Decile class is 40–45; $L = 40$, $C = 30$, $f = 24$; $D_2 = 40 + 5 \times \left(\frac{31.2 - 30}{24}\right) = 40 + 0.25 = 40.25$

For 5th Decile, 5th Decile class is 45–50; $L = 45$, $C = 54$, $f = 28$; $D_5 = 45 + 5 \times \left(\frac{78 - 54}{28}\right) = 45 + 4.29 = 49.29$

For 9th Decile, 9th Decile class is 60–65; $L = 60$, $C = 136$, $f = 14$; $D_9 = 60 + 5 \times \left(\frac{140.4 - 136}{14}\right) = 60 + 1.57 = 61.57$

****5th Decile is same as median**

Measures of Dispersion

Quartile deviation, $Q.D = \frac{(Q_3 - Q_1)}{2} = \frac{(56.83 - 42.25)}{2} = 7.29$

Variance, $\sigma^2 = \frac{\sum_{i=1}^{n} f_i x_i^2}{N} - \left(\frac{\sum_{i=1}^{n} f_i x_i}{N}\right)^2 = \frac{392575}{156} - \left(\frac{7690}{156}\right)^2 = 86.52$

Standard Deviation,

$$\sigma = \sqrt{\frac{\sum_{i=1}^{n} f_i x_i^2}{N} - \left(\frac{\sum_{i=1}^{n} f_i x_i}{N}\right)^2} = \sqrt{\frac{392575}{156} - \left(\frac{7690}{156}\right)^2} = \sqrt{86.52} = 9.30$$

Coefficient of variation $\text{C.V.} = \frac{\sigma}{\bar{x}} \times 100 = \frac{9.30}{49.29} \times 100 = 18.87$

First order central moment, $\mu_1 = 0$; Second order central moment, $\mu_2 =$ variance, $\sigma^2 = 86.52$

Third order central moment, $\mu_3 = \frac{\sum_{i=1}^{n} f_i (x_i - \bar{x})^3}{N} = \frac{-2080.54}{156} = -13.34$

Fourth order central moment, $\mu_4 = \frac{\sum_{i=1}^{n} f_i\left(x_i - \overline{x}\right)^4}{N} = \frac{24574218}{156} = 15752.7$

Measure of Skewness

Coefficient of skewness,

$$\beta_1 = \frac{\mu_3^2}{\mu_2^3} = \frac{(-13.34)^2}{(86.52)^3} = \frac{177.87}{647663.66} = 0.000225$$

$$\gamma_1 = \frac{\mu_3}{\mu_2^{3/2}} = \frac{-13.34}{(86.52)^{3/2}} = -0.0166$$

Measure of Kurtosis

Coefficient of kurtosis,

$$\beta_2 = \frac{\mu_4}{\mu_2^2} = \frac{15752.7}{(86.52)^2} = 2.104; \gamma_2 = \beta_2 - 3 = 2.104 - 3 = -0.896$$

Thus the series is negatively skewed and platykurtic.

Chapter 6

Probability

In some experiments the result can be predicted with certainty. For example, the bulb will stop glowing when switched off, the evolution of hydrogen gas when zinc reacts with dilute hydrochloric acid. Such phenomenon where the result can be predicted with certainty is known as deterministic phenomenon. But in some experiments the result cannot be predicted with certainty. For example, Sales of a firm, profit made by a company, yield of a crop, etc. Such phenomenon where the result cannot be predicted with certainty are known as probabilistic or random phenomenon. Such phenomena are frequently observed in economics, business, agriculture, social science, biology, etc., and in our day to day life.

The theory of probability is a very important branch of statistics which provides a numerical measure of uncertainty.

Some terms connected with probability

1. **Random experiment:** An experiment which when conducted repeatedly under essentially homogeneous & identical conditions does not give unique or certain results, but the result may be any one of the possible outcomes is called random experiment.

   For example, Yield of a particular crop in various plots of equal sizes under identical conditions and treatments.
2. **Trial:** Performing of a random experiment is called trial.
3. **Event:** Outcome or combination of outcomes of a trial is called and event.
4. **Favorable Event:** The outcome of a random experiment which results in the happening of the events is termed as favorable event. For example, in toss of three coins, the number of cases favorable to the event "getting one heads" is 3. The outcomes favourable to this event are {HTT, THT, TTH}.
5. **Exhaustive Events:** The total number of possible outcome of a random experiment. For example, in toss of three coins, the number of exhaustive cases is $2^3 = 8$, which are {HHH, HHT, HTH, THH, TTH, THT, HTT, TTT}.

6. **Mutually Exclusive Events:** Two or more events are said to be mutually exclusive or incompatible if the events can never occur simultaneously in the same trial. For example, in toss of three coins, the events of getting "two heads" and exactly "three heads" are mutually exclusive. But the events of "getting two heads" and "getting more than one head" are not mutually exclusive because if two heads occur then both the events will happen simultaneously.
7. **Complementary Events:** Two events are said to be complementary if they are mutually exclusive and exhaustive. For example, events E_1 of "getting odd numbers" and E_2 "getting even numbers" in throw of die. E_1= { 1, 3, 5}; E_2 = (2, 4, 6}
8. **Equally likely events:** Events having equal chances of occurrence or non-occurrence are called equally likely events. For example, in toss of an unbiased coin, the events of "getting head" and "getting tail" are equally likely events.
9. **Independent events:** Two or more events are said to be independent if the happening or non-happening of one event does not affect and is not affected by the happening or non-happening of the other event. For example, drawing of cards with replacement, the events of outcome of second throw and first throw of a die.
10. **Dependent Events:** Events which affects and/or is affected by the happening or non-happening of other events are called dependent events. For example, drawing of cards without replacement.
11. **Sample Space:** The set of all possible outcomes of a random experiment. It is denoted by S. Elements of the sample space are called Sample Points. For example, in toss of three coins, sample space (S) = {HHH, HHT, HTH, THH, THT, TTH, HTT, TTT}. Here number of sample points is 2^3 = 8. Number of elements or sample points in a sample space i.e., number of possible outcomes of a trial = (number of possible outcomes in one trial)^n, where n is the no. of trials.

Simple event or elementary event: Event having only one favourable outcome. E.g., In toss of 3 coins, the event of "getting 3 heads", "getting 3 tails" are all elementary events.

Compound event: Events having more than one favourable outcomes, for example, In toss of 3 coins, the event of "getting two heads" is a compound event.

Certain event or sure event: Event having all the possible no. of favourable outcomes. It contains all the sample points of sample space. Example, In throw of dice, the event of "getting less than 7" is a sure event.

Impossible event: Event having no favourable outcomes. It is a null event. It has no sample points. For example, In throw of dice the event of "getting more than 6". It is denoted as ϕ.

Definition of probability

Classical or mathematical definition

If a trial results in 'n' mutually exclusive, exhaustive & equally likely outcomes and 'm' of them are favourable to the happening of the event E, then the probability of happening of the event E, is given by the ratio of the number of favourable outcomes to the no. of possible outcomes. Denoted as $P(E)$ or $p = m/n$; Given by Laplace.

Statistical or empirical definition

If large number of trials are repeated under essentially homogeneous and identical conditions, then the limiting value of the ratio of the no. of times the event happens (m) to the no. of trials (n) as the no. of trials become indefinitely large i.e., $n \to \infty$ is called probability of the happening of event E.

P(E) or $p = \lim_{n\to\infty} m/n$, given by Van Mises.

Axiomatic definition

This is the modern definition of probability given by Kolmogorov. Given a sample space 'S' of a random experiment, the probability of occurrence of any event E is defined as a set function $P(E)$ satisfying the following three axioms:

i. $P(E) \geq 0$

ii. $P(E1 \cup E2 \cup \ldots \cup \ldots En) = P(E1) + P(E2) + \ldots + P(En)$

i.e., $P\left(\bigcup_{i=1}^{n} E_i\right) = \sum_{i=1}^{n} P(E_i)$, if $E_1, E_2, \ldots, E_n$ are mutually exclusive.

iii. $P(S) = 1$; $S \to$ sample space, are mutually exclusive.

Properties of probability

1. Probability of impossible event is zero i.e., $P(\phi) = 0$
2. Probability of complementary event of A is given by $P(A') = 1 - P(A)$
3. $P(A \cap B) = P(A) - P(A \cap B)$
4. $P(A \cap B) = P(B) - P(A \cap B)$
5. Addition law of probability

For two events, $P(A_1 \cup A_2) = P(A_1) + P(A_2) - P(A_1 \cap A_2)$

For three events, $P\left(\bigcup_{i=1}^{3} A_i\right) = \sum_{i=1}^{3} P(A_i) - \sum_{i=1}^{2}\sum_{j=2}^{3} P(A_i \cap A_j) - P\left(\bigcap_{i=1}^{3} A_i\right)$

For four events:

$$P\left(\bigcup_{i=1}^{4} A_i\right) = \sum_{i=1}^{4} P(A_i) - \sum_{i=1}^{3}\sum_{j=2}^{4} P(A_i \cap A_j) + \sum_{i=1}^{2}\sum_{j=2}^{3}\sum_{k=3}^{4} P(A_i \cap A_j \cap A_k) - P\left(\bigcap_{i=1}^{4} A_i\right)$$

For n events:

$$P\left(\bigcup_{i=1}^{n} A_i\right) = \sum_{i=1}^{n} P(A_i) - \sum_{i=1}^{n-1}\sum_{j=2}^{n} P(A_i \cap A_j) + \sum_{i=1}^{n-2}\sum_{j=2}^{n-1}\sum_{k=3}^{n} P(A_i \cap A_j \cap A_k) - \sum_{i=1}^{n-3}\sum_{j=2}^{n-2}\sum_{k=3}^{n-1}\sum_{l=4}^{n} P(A_i \cap A_j \cap A_k \cap A_l) + (-1)^{n+1} P\left(\bigcap_{i=1}^{n} A_i\right)$$

For mutually exclusive events, $P\left(\bigcup_{i=1}^{n} A_i\right) = \sum_{i=1}^{n} P(A_i)$

6. Multiplication law of probability
   $P(A_1 \cap A_2) = P(A_1).P(A_2/A_1)$; $P(A_1) > 0$ or
   $P(A_1 \cap A_2) = P(A_2).P(A_1/A_2)$; $P(A_2) > 0$
   $P(A_1)$ = unconditional probability of event A_1;
   $P(A_2)$ = unconditional probability of event A_2
   $P(A_1/A_2)$ = conditional probability of event A_1 given that event A_2 has already occurred.
   $P(A_2/A_1)$ = conditional probability of event A_2 given that event A_1 has already occurred.
   $P(A_1 \cap A_2 \cap A_3) = P(A_1).\ P(A_2/A_1).P(A_3/A_1 \cap A_2)$
   $P(A_3/A_1 \cap A_2)$- conditional probability of occurrence of event A_3 given that events A_1 and A_2 has already occurred.
   For n events, $P(A_1 \cap A_2 \cap A_3 \cap \ldots \cap A_n) = P(A_1)\cdot P(A_2/A_1)\cdot P(A_3/A_1 \cap A_3) \ldots P(A_n/A_1 \cap A_2 \cap \ldots \cap A_{n-1})$
   Independent events: An event A_2 is said to be independent of event A_1, if the conditional probability of event A_2 given A_1 [i.e., $P(A_2/A_1)$] is equal to unconditional probability of event A_2 [$P(A_2)$]i.e., $P(A_2/A_1) = P(A_2)$
   Similarly an event A_1 is said to be independent of event A_2, if the conditional probability of event A_1 given A_2 [i.e., $P(A_1/A_2)$] is equal to unconditional probability of event A_1 [$P(A_1)$] i.e., $P(A_1/A_2) = P(A_1)$.
   If A_1 and A_2 are independent, then the multiplicative law becomes
   $P(A_1 \cap A_2) = P(A_1)P(A_2)$
   For three events, it becomes $P(A_1 \cap A_2 \cap A_3) = P(A_1)P(A_2)P(A_3)$
   For n events, it becomes
   $P(A_1 \cap A_2 \cap A_3 \cap \ldots \cap A_n) = P(A_1)P(A_2)P(A_3)\ldots P(A_n)$
   i.e., $P\left(\bigcap_{i=1}^{n} A_i\right) = P\left(\prod_{i=1}^{n} A_i\right)$

Numerical Problems

1. A Problem in statistics is given to three students named as *A*, *B* and *C*. The probability that *A*, *B* and *C* can solve the problem are respectively 1/3, 1/4 and 2/5. If they try independently, then find the probability that (i) Problem can be solved (ii) Problem cannot be solved (iii) can be solved by at least any two of the students (iv) can be solved by at most any two.

Solution

Let *A* be the event that the problem is solved by *A*, *B* be the event that the problem is solved by *B*, *C* be the event that the problem is solved by *C*.

Given that, $P(A) = 1/3$, $P(B) = 1/4$ and $P(C) = 2/5$. Events *A*, *B* and *C* are independent.

Let *E* be the event that the problem can be solved. The event *E* can materialise in 3 ways: E_1, E_2 and E_3. E_1 – Event that the problem can be solved by any one; E_2 – Event that the problem can be solved by any two; E_3 – Event that the problem can be solved by any three. We have, $E = E_1 \cup E_2 \cup E_3$.

Taking probability on both sides, we get

$$P(E) = P(E_1 \cup E_2 \cup E_3)$$
$$= P(E_1) + P(E_2) + P(E_3) \; (\because E_1, E_2 \text{ and } E_3 \text{ are mutually exclusive})$$

Event E_1 can materialise in 3 ways E_{1A} (Event that only *A* can solve), E_{1B} (Event that only *B* can solve) , E_{1C} (Event that only *C* can solve).

We have, $E_1 = E_{1A} \cup E_{1B} \cup E_{1C}$

Taking probability on both sides, we get

$$P(E_1) = P(E_{1A} \cup E_{1B} \cup E_{1C})$$
$$= P(E_{1A}) + P(E_{1B}) + P(E_{1C})$$

($\because$ the events E_{1A}, E_{1B} and E_{1C} are mutually exclusive)

$$= P(A \cap B' \cap C') + P(A' \cap B \cap C') + P(A' \cap B' \cap C)$$

$$\{\text{where, } P(A') = 1 - P(A);\; P(B') = 1 - P(B);\; P(C') = 1 - P(C)\}$$

$$= P(A).P(B').P(C') + P(A').P(B).P(C') + P(A').P(B').P(C)$$

(Since events *A*, *B* and *C* are independent)

$$= P(A)\,\{1 - P(B)\}\{1 - P(C)\} + \{1 - P(A)\}P(B)\,\{1 - P(C)\} + \{1 - P(A)\{1 - P(B)\}P(C)$$

$$= \frac{1}{3} \times \left\{1 - \frac{1}{4}\right\} \times \left\{1 - \frac{2}{5}\right\} + \left\{1 - \frac{1}{3}\right\} \times \frac{1}{4} \times \left\{1 - \frac{2}{5}\right\} + \left\{1 - \frac{1}{3}\right\} \times \left\{1 - \frac{1}{4}\right\} \times \frac{2}{5}$$

$$= \frac{1}{3} \times \frac{3}{4} \times \frac{3}{5} + \frac{2}{3} \times \frac{1}{4} \times \frac{3}{5} + \frac{2}{3} \times \frac{3}{4} \times \frac{2}{5} = \frac{3}{20} + \frac{2}{20} + \frac{4}{20} = \frac{9}{20}$$

Event E_2 can materialise in 3 ways E_{2AB} (Event that only A and B can solve), E_{2BC} (Event that only B and C can solve), E_{2AC} (Event that only A and C can solve). We have, $E_2 = E_{2AB} \cup E_{2BC} \cup E_{2AC}$

Taking probability on both sides, we get

$$P(E_2) = P(E_{2AB} \cup \text{U } E_{2BC} \cup E_{2AC})$$

$$= P(E_{2AB}) + P(E_{2BC}) + P(E_{2AC})$$

($\because$ the events E_{2AB}, E_{2BC} and E_{2AC} are mutually exclusive)

$= P(A \cap B \cap C') + P(A' \cap B \cap C) + P(A \cap B' \cap C)$

$= P(A).P(B).P(C') + P(A').P(B).P(C) + P(A).P(B').P(C)$

($\because$ events A, B and C are Mutually Exclusive)

$= P(A)P(B)\{1 - P(C)\} + \{1 - P(A)\}P(B)\,P(C) + P(A)\{1 - P(B)\}P(C)$

$$= \frac{1}{3} \times \frac{1}{4} \times \left\{1 - \frac{2}{5}\right\} + \left\{1 - \frac{1}{3}\right\} \times \frac{1}{4} \times \frac{2}{5} + \frac{1}{3}\left\{1 - \frac{1}{4}\right\}\frac{2}{5}$$

$$= \frac{1}{3} \times \frac{1}{4} \times \frac{3}{5} + \frac{2}{3} \times \frac{1}{4} \times \frac{2}{5} + \frac{1}{3} \times \frac{3}{4} \times \frac{2}{5}$$

$$= \frac{1}{20} + \frac{1}{15} + \frac{1}{10}$$

$$= \frac{13}{60}$$

$$P(E_3) = P(A \cap B \cap C) = P(A).P(B).P(C) = \frac{1}{3} \times \frac{1}{4} \times \frac{2}{5} = \frac{2}{60}$$

(i) Thus, from equation 1, $P(E) = \frac{9}{20} + \frac{13}{60} + \frac{2}{60} = \frac{42}{60} = 0.7$

(ii) P(problem cannot be solved) $= P(E') = P(A' \cap B' \cap C') = P(A')P(B')$

$$P(C') = \left\{1 - \frac{1}{3}\right\}\left\{1 - \frac{1}{4}\right\}\left\{1 - \frac{2}{5}\right\} = \frac{2}{3} \times \frac{3}{4} \times \frac{3}{5} = \frac{6}{20} = 0.3$$

(iii) P (at least any two) $P(E_2) + P(E_3) = \frac{13}{60} + \frac{2}{60} = \frac{15}{60} = 0.25$

(iv) P (at most any two can solve) = P(solved by none) + P(solved by any one) + P(solved by any two)

$$= P(E') + P(E_1) + P(E_2) = \frac{6}{20} + \frac{27}{60} + \frac{13}{60} = \frac{58}{60}$$

or, P(at most any two can solve) $= 1 - P(E_3) = 1 - \frac{2}{60} = \frac{58}{60}$

2. There are 6 mangoes, 3 coconuts and 5 guavas in a bag. Three fruits are drawn at random from the bag. Find the probability that (i) 2 mangoes and 1 coconut is drawn (ii) one guava is drawn (iii) all fruits are of one kind (iv) all fruits are of different kinds.

Solution

Total number of fruits in the bag = 6 + 3 + 5 = 14

Number of ways in which three fruits can be drawn randomly from a bag containing 14 fruits = $C(14, 3)$

Thus, $n(S) = C(14.3)$, where S denotes the Sample space.

(i) Let E be the event of drawing 2 mangoes and 1 coconut. $n(E) = C(5, 2) \times C(4, 1)$

$$\therefore P(E) = \frac{C(6, 2) \times C(3, 1)}{C(14, 3)} = \frac{15 \times 3}{364} = \frac{45}{364}$$

(ii) Let E be the event of drawing at least one guava. Event E can materialise in 3 ways:

E_1 – 1 guava and 2 mangoes, E_2 – 1 guava and 2 coconuts, E_3 = 1 guava, 1 mango and 1 coconut

$E = E_1 \cup E_2 \cup E_3 \Rightarrow P(E) = P(E_1 \cup E_2 \cup E_3) = P(E_1) + P(E_2) + P(E_3)$

($\because E_1$, E_2 and E_3 are mutually exclusive)

$$P(E_1) = \frac{C(5,1) \times C(6,2)}{C(14,3)} = \frac{5 \times 15}{364} = \frac{75}{364}$$

$$P(E_2) = \frac{C(5,1) \times C(3,2)}{C(14,3)} = \frac{5 \times 3}{364} = \frac{15}{364}$$

$$P(E_3) = \frac{C(5,1) \times C(6,1) \times C(3,1)}{C(14,3)} = \frac{5 \times 6 \times 3}{364} = \frac{90}{364}$$

$$P(E) = \frac{75}{364} + \frac{15}{364} + \frac{90}{364} = \frac{180}{364}$$

(iii) Let E be the event that all fruits are of one kind. E can materialise in 3 ways:

E_1 – all are mangoes, E_2 – all are coconuts, E_3 = all are guavas

$$P(E_1) = \frac{C(6,3)}{C(14,3)} = \frac{20}{364}$$

$$P(E_2) = \frac{C(3,3)}{C(14,3)} = \frac{1}{364}$$

$$P(E_3) = \frac{C(5,3)}{C(14,3)} = \frac{10}{364}$$

$E = E_1 \cup E_2 \cup E_3 \Rightarrow P(E) = P(E_1 \cup E_2 \cup E_3) = P(E_1) + P(E_2) + P(E_3)$
($\because E_1, E_2$ and E_3 are mutually exclusive)

$$\therefore\ P(E) = \frac{20}{364} + \frac{1}{364} + \frac{10}{364} = \frac{31}{364}$$

(iv) Let E be the event that all fruits are of different kinds.

$$P(E) = \frac{C(6,1) \times C(5,1) \times C(3,1)}{C(14,3)} = \frac{90}{364}$$

3. Find the probability that a non-leap year has (i) 53 Mondays (ii) 54 Mondays (iii) 52 Mondays

Solution

(i) In a non-leap year, 365 days = 7 × 52 weeks + 1 days; If the remaining 1 day will be Monday then there will be 53 Mondays.

Sample space, S = {Sun, Mon, Tue, Wed, Thu, Fri, Sat}; Let E be the event that the remaining one day is a Monday. E = {Mon};

Thus, $n(E) = 1$, $n(S) = 7$. $P(E) = \frac{n(E)}{n(S)} = \frac{1}{7}$

(ii) P(54 Mondays) = 0, an impossible event; P(52 Mondays) = 1, a sure event

4. Find the probability that a leap year has (i) 53 Sundays (ii) 54 Sundays (iii) 52 Sundays

Solution

In a leap year, 366 days = 7 × 52 weeks + 2 days; If in the remaining 2 days, any one day will be Sunday then there will be 53 Sundays. Sample space, S = {(sun, mon), (mon, tue), (tue, wed), (wed, thu), (thu, fri), (fri, sat), (sat, sun)}; Let E be the event that in the remaining 2 days any one day will be Sunday.

E = {(sun, mon), (sat, sun)}; Thus, $n(E) = 2$, $n(S) = 7$. $P(E) = \frac{n(E)}{n(S)} = \frac{2}{7}$

(ii) P(54 Sundays) = 0, an impossible event; P(52 Sundays) = 1, a sure event

5. $P(A \cap B) = 0.6$, $P(B) = 0.9$, $P(A/B) = ?$

Solution $P(A/B) = \frac{P(AB)}{P(B)} = \frac{0.6}{0.9} = 0.67$

6. $P(A) = 0.6$, $P(B) = 0.4$, Find $P(A \cup B)$ if A and B are independent.

Solution

$P(A \cup B) = P(A) + P(B) - P(A \cap B) = P(A) + P(B) - P(A).P(B)$

($\because A$ and B are independent)

$= 0.6 + 0.4 - 0.6 \times 0.4 = 0.76$

7. $P(A) = 0.6$, $P(B) = 0.5$, $P(A \cup B) = 0.9$, find $P(A \cap B)$

Solution: $P(A \cup B) = P(A) + P(B) - P(A \cap B)$

$\Rightarrow P(A \cap B) = P(A) + P(B) - P(A \cup B)$

$= 0.6 + 0.5 - 0.9 = 0.20.$

8. $P(A)=1/2$; $P(B)=2/5$; $P(A \cap B) = 1/3$. Find (i) $P(\bar{A} \cap \bar{B})$ (ii) $P(\bar{A} \cup \bar{B})$

(iii) $P(A \cap \bar{B})$ (iv) $P(\bar{A} \cap B)$

Solution: $P(A \cup B) = P(A) + P(B) - P(A \cap B) = \frac{1}{2} + \frac{2}{5} - \frac{1}{3} = \frac{15+12-10}{30} = \frac{17}{30}$

(i) $P(\bar{A} \cap \bar{B}) = P(\overline{A \cup B}) = 1 - P(A \cup B) = 1 - \frac{17}{30} = \frac{13}{30}$

(ii) $P(\bar{A} \cup \bar{B}) = P(\overline{A \cap B}) = 1 - P(A \cap B) = 1 - \frac{1}{3} = \frac{2}{3}$

(iii) $P(A \cap \bar{B}) = P(A) - P(A \cap B) = \frac{1}{2} - \frac{1}{3} = \frac{1}{6}$

(iv) $P(\bar{A} \cap B) = P(B) - P(A \cap B) = \frac{2}{5} - \frac{1}{3} = \frac{1}{15}$

Chapter 7

Probability Distribution

Random variable: A random variable is a variable that assumes numerical values associated with the events of a random experiment whose outcomes cannot be predicted with certainty. For example, Height of rice plants in a field; seed weight; No. of tillers per rice plant.

A random variable is of two types:-

1. **Discrete random variable:** The random variable which takes only countable values i.e., integral values. For Example, Number of tillers per rice plant. Here there can be finite no. of values between any two values.
2. **Continuous random variables:** The random variable which takes values within intervals. These values cannot be counted but can only be measured. Here there can be infinite no. of values between any two values for example, Height of rice plants in field; seed weight.

Probability distribution: The set of all distinct values (either point values or intervals) of the random variable along with their corresponding probabilities is called probability distribution.

Probability distribution of a discrete random variable is called discrete probability distribution, where as that of continuous variable is called continuous probability distribution.

Example of discrete probability distribution: If X is a discrete random variable denoting the no. of tillers in rice plant. Suppose we select 600 rice plants out of which 60 plants have 20 tillers/plant, 120 plants have 22 tillers/plant, 180 plants have 24 tillers/plant and 240 plants have 28 tillers/plant, then the prob. distribution of random variable X is given by

X_i	20	22	24	28
$P(x_i)$	$\frac{60}{600}=0.1$	$\frac{120}{600}=0.2$	$\frac{180}{600}=0.3$	$\frac{240}{600}=0.4$

Here $p(x_i)$ is the probability mass function of random variable X.

Example of continuous prob. Distribution: If X is a continuous random variable denoting the height of rice plant and we select 900 rice plants. Suppose out of 900 rice plants: 90 plant have height in the interval of

80–90 cm: 135 plants are in between 90–100 cm; 225 plants are in between 100–110 cm; remaining 450 plants have height in the range of 110–120 cm. Prob. Distribution of random variable X is given by:

Height (in cm) (x_i)	80–90	90–100	100–110	110–120
Probability $P(a < x < b)$	$P(80 \le x_i \le 90)$ $= \frac{90}{900} = 0.1$	$P(90 \le x_i \le 100)$ $= \frac{135}{900} = 0.15$	$P(100 \le x_i \le 110)$ $= \frac{225}{900} = 0.25$	$P(110 \le x_i \le 120)$ $= \frac{225}{900} = 0.50$

$p(x)$ is called prob. density function of random variable X.

Probability function: The probability associated with the values of a random variable is a function of the random variable and is called probability function.

Probability mass function(p.m.f): If 'X' is a discrete random variable taking values $x_1, x_2, \ldots, x_n$ then with each value of x_i(I = 1, 2, …,n) we associate a number $p(x_i)$ or p_i or $P(X = x_i)$ such that $0 \le p(x) \le 1$ and $\sum_x p(x) = 1$, then the function $p(x_i)$ is called the p.m.f. of the random variable X.

Probability density function(p.d.f): If 'X' is a continuous random variable and $f(x)$ is a continuous function of x, such that $f(x)dx$ gives the probability of event that x lies in the interval $x - \frac{dx}{2}$ and $x + \frac{dx}{2}$ i.e., $P\left(x - \frac{dx}{2} d \times x + \frac{dx}{2}\right)$, then the function $f(x)$ is called p.d.f. of the random variable X, provided that $0 \le f(x) \le 1$ and $\int_x f(x)dx = 1$ p.m.f at a particular point may or may not be zero, but p.d.f. at a particular point is always zero.

Mathematical expectation: Mathematical expectation or expected value of a random variable is the sum of the products of the values of the variable and their corresponding probabilities. It is denoted as $E(X)$.

Thus, $E(X) = \sum_{i=1}^{n} x_i p(x_i)$ if X is a discrete random variable

$E(X) = \int_x xf(x)dx,$ if X is a continuous random variable.

Expected value of random variable X i.e., $E(X)$ is same as mean of the random variable X.

Properties of Mathematical Expectations

1. $E(X_1 + X_2 + \cdots + X_n) = E(X_1) + E(X_2) + \cdots + E(X_n)$

   i.e., $E\left(\sum_{i=1}^{n} X_i\right) = \sum_{i=1}^{n} E(X_i)$

2. $E(X_1 \cdot X_2 \cdot X_3 \cdots X_n) = E(X_1).\, E(X_2) \cdots E(X_n)$ i.e., $E\left(\prod_{i=1}^{n} X_i\right) = \prod_{i=1}^{n} E(X_i)$, provided X_i's are independent.
3. $E(a) = a$; 'a' is a constant.
4. $E(aX) = aE(X)$,
5. $E(aX \pm b) = aE(X) \pm b$, a, b e constant.
6. $E(a_1X_1 + a_2X_2 + \cdots + a_nX_n) = a_1E(X_1) + a_2E(X_2) + \cdots + a_nE(X_n)$

   i.e., $E\left(\sum_{i=1}^{n} a_i X_i\right) = \sum_{i=1}^{n} E(a_i X_i)$
7. $\text{Var}(aX) = a^2\text{Var}(X)$; a is any constant
8. $\text{Var}(aX \pm b) = a^2\text{var}(X)$, a, b are constants.

   Variance in terms of expectation is written as

   $V(X) = E[\{X - E(X)\}^2] = E(X)^2 - \{E(X)\}^2$

   Covariance in terms of expectation is

   $\text{COV}(X, Y) = E[\{X - E(X)\}\{Y - E(Y)\}] = E(XY) - E(X)\,E(Y)$

   $E(XY) = \sum_x \sum_y xy\, p(x, y)$; If X and Y are discrete random variables. $p(x, y)$ is the joint probability of the random variables X and Y.

   $E(XY) = \int_x \int_y x.y f(x, y)\,dx dy$. If X and Y are continuous random variables. $f(x, y)$ is the joint probability of the random variables X and Y.

Numerical Problems

1. If X and Y are independent, then prove that $\text{COV}(X, Y) = 0$.

   Solution: If X and Y are independent then $E(XY) = E(X)\,.\,E(Y)$

   So, $\text{COV}(X, Y) = E(XY) - E(X)\,.\,E(Y) = E(X)\,.\,E(Y) - E(X)\,.\,E(Y) = 0$
2. For the following probability distribution of random variable X, find Mean and Variance of X:

x_i	6	5	7	8	9
$p(x_i)$	0.2	0.1	0.25	0.15	0.3

Solution

x_i	6	5	7	8	9	
$p(x_i)$	0.2	0.1	0.25	0.15	0.3	
$x_i \times p(x_i)$	1.2	0.5	1.75	1.2	2.7	$\sum_{i=1}^{5} x_i p(x_i) = 7.35$
$x_i^2 \cdot p(x_i)$	7.2	2.5	12.25	9.6	24.3	$\sum_{i=1}^{5} x_i^2 p(x_i) = 55.85$

Mean, $E(X) = \sum_{i=1}^{5} x_i p(x_i) = 7.35; E(X^2) = \sum_{i=1}^{5} x_i^2 p(x_i) = 55.85$

So, Variance, $V(X) = E(X^2) - \{E(X)\}^2$

$= 55.85 - (7.35)^2 = 55.85 - 54.0225 = 1.8275$

3. If $E(X) = 60$ and $E(Y) = 10$ then find (i) $E(3X - 7)$ (ii) $E(5Y + 9)$ (iii) $E(8X)$ (iv) $E(7)$

Solution

(i) $E(3X - 7) = 3E(X) - 7 = 3 \times 60 - 7 = 173$

(ii) $E(5Y + 9) = 5E(Y) + 9 = 5 \times 10 + 9 = 59$

(iii) $E(8X) = 8E(X) = 8 \times 60 = 480$

(iv) $E(7) = 7$

4. If $V(X) = 16$, $V(Y) = 12$, $COV(X, Y) = 8$, then find

(i) $V(2X - 8)$

(ii) $V(X - Y)$

(iii) $V(X + Y)$

(iv) $V(2X - 3Y)$

(v) $V(X + 6Y)$

(vi) $V(3X + 6)$

Solution

(i) $V(2X - 8) = 2^2 V(X) = 4 \times 16 = 64$

(ii) $V(X - Y) = V(X) + V(Y) - 2COV(X, Y) = 16 + 12 - 2 \times 8 = 12$

(iii) $V(X + Y) = V(X) + V(Y) + 2COV(X, Y) = 16 + 12 + 2 \times 8 = 44$

(iv) $V(2X - 3Y) = 2^2V(X) + 3^2V(Y) - 2x2x3COV(X, Y)$

$= 4 \times 16 + 9 \times 12 - 12 \times 8 = 64 + 108 - 96 = 76$

(v) $V(X + 6Y) = V(X) + 6^2V(Y) - 2x1x6\ COV(X, Y)$

$= 16 + 36 \times 12 - 12 \times 8 = 16 + 432 - 96 = 352$

(vi) $V(3X + 6) = 3^2V(X) = 9 \times 16 = 144$

Two important Discrete Probability Distributions are: **(A) Binomial Distribution** and **(B) Poisson Distribution**

(A) Binomial Distribution

Definition: A random variable X is said to follow binomial distribution if it assumes non- negative integral values 0, 1, 2,...,n and its p.m.f. is given by

$$p(x) = {}^nC_x p^x q^{n-x}; \text{ for } x = 0, 1, 2\ldots,n$$
$$= 0 \quad ; \text{otherwise}$$

Where n is the number of trials;

p is the probability of success in any trial;

q is the probability of failure in any trial; ($q + p = 1$)

x is number of success in 'n' trials

n and p are called parameters of the Binomial distribution.

n is also called as degree of Binomial Distribution.

If X follows Binomial Distribution with parameters n and p, then symbolically, it is written as $X\sim B(n, p)$. Here X is called as a "Binomial variate".

Physical conditions for Binomial distribution

1. The number of trials, n, should be finite.
2. Each trial must result in two mutually exclusive and exhaustive outcomes, i.e., success and failure.
3. The trials must be independent of each other.
4. The probability of success in each trial should be same.

Examples of random variable that follows binomial distribution: Outcomes of tossing a coin, Outcomes of throwing a dice, Outcomes of drawing cards with replacement, no. of seeds germinated in different petridishes, Disease incidence in plants of a field, etc.

Uses of binomial distribution

1. Binomial distribution is used to find out the probability that a particular amount of seeds will germinate out of a given number of seeds, provided that germination % of seeds are known.
2. Binomial distribution serves as a parent distribution for other distributions like Poisson distribution, Normal distribution, etc.

Properties of binomial distribution

1. Mean = np and variance = npq. As $q < 1$; so, mean > variance. This is the important property of Binomial distribution.
2. Mode: If $(n + 1)p$ is not an integer, then the Binomial Distribution has only one mode. In this case, the integral part of $(n + 1)p$ gives the mode. If $(n + 1)p$ is an integer, than the Binomial Distribution is bimodal i.e., it has two modes. The two modes are $(n + 1)p$ and $\{(n + 1)p - 1\}$

(B) Poisson Distribution

It is a limiting case of binomial distribution i.e., binomial distribution tends to Poisson Distribution under the following three conditions:

1. The number of trials, n, should be indefinitely large i.e., $n \to \infty$
2. The probability of success 'p' for each trial is indefinitely small.
3. $np = \lambda$ (say), should be finite, where λ is a positive real number.

Some practical situations in which Poisson Distribution can be used

1. The occurrence of rainfall on a particular day.
2. The mortality of plants in a plot during the crop period, by a rare disease.
3. Number of weed plants in different plots, after weedicide have been applied.
4. Number of insects caught in a trap in one minute.

Definition: A random variable 'x' is said to follow Poisson Distribution if it takes non-negative integral values and its p.m.f is given by:

$$p(x) = (e^{-\lambda}\lambda^{x})/x!;\ x = 0, 1, 2, \ldots.$$
$$= 0 \qquad ;\ \text{otherwise}$$

Where λ is the parameter of Poisson Distribution, x is the number of success and $e = 2.71$.

Symbolically, if X is a Poisson variate with parameter λ, then it is denoted as: $X \sim P(\lambda)$.

Characteristics/Properties of Poisson Distribution

1. Mean = λ; Variance = λ; i.e., mean and variance of Poisson Distribution are equal (important property)
2. Mode – If λ is not an integer, then Poisson Distribution is unimodal. The mode is given by integral part of λ. if λ is an integer, then there are two modes which are $(\lambda - 1)$ and λ.

The most important Continuous Probability Distribution is Normal Distribution

Normal Distribution

Definition: A random variable X is said to follow normal distribution with parameters μ (mean) and σ^2 (variance) if its p.d.f is given by:

$$f(x) = \frac{1}{\sqrt{2\pi\sigma^2}} e^{-\frac{1}{2}\left(\frac{x-\mu}{\sigma}\right)^2};\ -\infty < x < \infty;\ -\infty < \mu < \infty;\ \sigma > 0$$
$$= 0\ ;\ \text{otherwise.}$$

μ (mean) and σ^2 (variance) are called the parameters of normal distribution.

If X follows normal distribution with parameters μ and σ^2, symbolically it is written as:

$X \sim N(\mu, \sigma^2)$. Here X is called as a normal variate.

Standard Normal Variate:

If $X \sim N(\mu, \sigma^2)$, then $z = \frac{x-\mu}{\sigma}$ is called Standard Normal Variate.

p.d.f of standard Normal variate is given by

$$f(z) = \frac{1}{\sqrt{2\pi}} e^{-\frac{1}{2}z^2}; -\infty < Z < \infty = 0; \text{ otherwise.}$$

mean of $Z, E(z) = 0, \text{var}(Z) = \sigma_z^2 = 1$

Definition of standard normal distribution

A random variable 'z' is said to follow standard normal distribution if its p.d.f. is given by $f(z) = \frac{1}{\sqrt{2\pi}} e^{-\frac{1}{2}z^2}; -\infty < z < \infty = 0$; otherwise.

Normal variate: The random variable X which taken continuous values and whose p.d.f. is given by

$$f(x) = \frac{1}{\sqrt{2\pi\sigma^2}} e^{-\frac{1}{2}\left(\frac{x-\mu}{\sigma}\right)^2}; -\infty \le x \le \infty; < \mu < \infty; \sigma > 0$$

$= 0$; otherwise is called Normal Variate.

Standard Normal Variate: The random variable Z which takes continuous values and whose p.d.f. is given by

$$f(z) = \frac{1}{\sqrt{2\pi}} e^{-\frac{1}{2}z^2}; -\infty < z < \infty$$

$= 0$; otherwise.

Normal Distribution: is a limiting form of Binomial Distribution i.e., Binomial Distribution tends to Normal Distribution under the following conditions:

1. n, the no. of trials is indefinitely large i.e., $n \to \infty$
2. Neither p nor q is very small.

Normal Probability Curve: The probability curve obtained by plotting x against $f(x)$ where $X \sim N(\mu, \sigma^2)$ and $f(x)$ is given by

$$f(x) = \frac{1}{\sigma\sqrt{2\pi}} e^{-\frac{1}{2}\left(\frac{x-\mu}{\sigma}\right)^2} \quad x \to$$

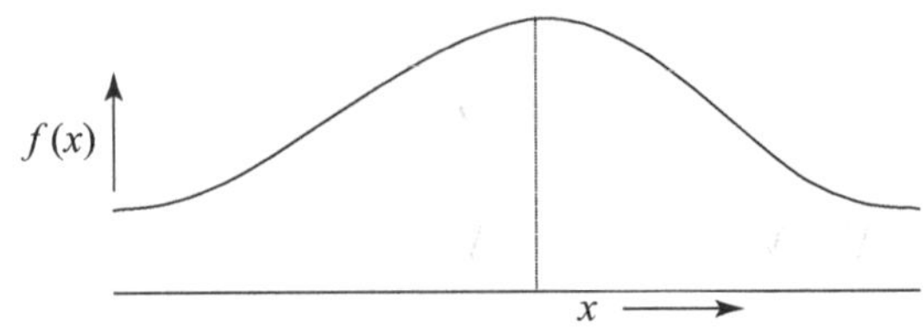

Standard Normal Probability Curve The prob. curve obtained by Plotting z against $f(z)$. Where $z \sim N(0, 1)$ and $f(z)$ is given by

$$f(Z) = \frac{1}{\sqrt{2\pi}} e^{-\frac{1}{2}z^2} \quad z \rightarrow$$

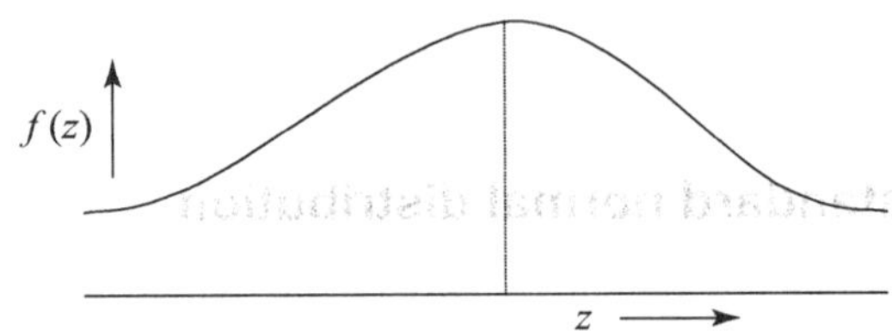

Characteristics/Properties of Normal distribution or Normal probability curve

1. The normal prob. curve is bell shaped and symmetrical about $x = \mu$
2. Mean, median & mode coincide (i.e., mean = median = mode).
3. The maximum probability occurs at the point $x = \mu$ and is given by

$$[f(x)]_{\max} = \frac{1}{\sigma\sqrt{2\pi}}$$

4. No portion of the curve lies below x-axis because $f(x)$ being the probability cannot be negative.
5. x-axis is asymptote to the curve i.e., the curve touches x-axis at $x = \pm\infty$
6. Area property (Important Property) $\rightarrow$ The probability that x lies within the interval a and b i.e., $P(a < x < b)$ is given by the area under the normal probability curve x-axis, ordinates at $x = a$ and $x = b$ i.e.,

$P(a < x < b) = \int_a^b f(x)dx =$ area of the shaded region.

$P(\mu - \sigma < x < \mu + \sigma) = P(-1 < z < 1) = 0.6826$

$P(\mu - 2\sigma < x < \mu + 2\sigma) = P(-2 < z < 2) = 0.9544$

$P(\mu - 3\sigma < x < \mu + 3\sigma) = P(-3 < z < 3) = 0.9973$

$P(\mu - 1.96\sigma < x < \mu + 1.96\sigma) = P(-1.96 < z < 1.96) = 0.95$

$P(\mu - 2.586 < x < \mu + 2.586) = P(-2.58 < z < 2.58) = 0.99$

$P(\mu - 2.33\sigma < x < \mu + 2.33\sigma) = P(-2.33 < x < 2.33) = 0.98$

$P(\mu - 1.64\sigma < x < \mu + 1.64\sigma) = P(-1.64 < x < 1.64) = 0.90$

Uses of Normal Distribution

1. Most of the distributions occurring in practice, e.g., Binomial Distribution, Poisson Distribution , etc., can be approximated by Normal Distribution e.g., student's *t*-test, Snedecor's *F*-test, x^2 – distribution, etc. tend to normality for large samples.
2. The area property of normal distribution forms the basis of entire large sample theory.

The entire theory of small sample tests, such as *t*-test, *F*-test, x^2 test, etc. is based on the fundamental assumption that the parent populations from which the samples are drawn follow normal distribution.

Numerical Problems

1. It is found that there is chance of finding 60 diseased plants out of every 300 coconut plants. Out of 12 selected plants, find the probability of getting (i) at least 2 diseased plants (ii) at most 2 diseased plants (iii) exactly 2 diseased plants.

Solution

Let the random variable X denotes the number of diseased plants. Number of trials, $n = 12$;
Probability that a plant will be diseased, $p = 60/300 = 0.20$.
Thus, $X \sim B(12, 0.20)$.
(i) P (at least 2 diseased plants) i.e.,

$$P(X \geq 2) = \sum_{x=2}^{12} C(12,x)(0.20)^x (0.80)^{12-x}$$

$$= 1 - \sum_{x=0}^{1} C(12,x)(0.20)^x (0.80)^{12-x}$$

$= 1 - \{C(12,0)\,(0.20)^0\,(0.780)^{12-0} + C(12,1)\,(0.20)^1(0.80)^{12-1}\}$
$= 1 - \{0.80^{12} + 12 \times 0.20 \times (0.80)^{11}\} = 1 - 0.275 = 0.725$

(ii)P(atmost2diseasedplants)i.e., $P(X \leq 2) = \sum_{x=0}^{2} C(12,x)(0.20)^x (0.80)^{12-x}$

$= C(12, 0)\,(0.20)^0\,(0.80)^{12-0} + C(12, 1)\,(0.20)^1(0.80)^{12-1}$
$+ C(12, 2)\,(0.20)^2(0.80)^{12-2}$
$= \{(0.80)^{12} + 12 \times 0.20 \times (0.80)^{11} + 66 \times (0.20)^2 \times (0.80)^{10} = 0.558$
(iii) P(exactly 2 diseased plants) i.e., $P(X = 2) = C(12, 2)\,(0.20)^2(0.80)^{122} = 66 \times (0.20)^2 \times (0.80)^{10} = 0.283$

2. There is chance of getting a diseased plant per 1000 plants in a field where spraying of appropriate fungicide has been done. Find the probability of getting 3 diseased plants in the entire field of 10000 plants.

Solution

Let the random variable X denotes the number of diseased plants in the field.

Here prob. of getting success in a single trial, $p = 1/1000 = 0.001$; number of trials, $n = 10000$.

Thus, $np = \lambda = 10000 \times 0.001 = 10$

Since n is indefinitely large and p is very small but np is a finite number, $X \sim P(10)$

The prob. of getting 3 diseased plants, $p(3) = \dfrac{e^{-\lambda}\lambda^3}{3!} = \dfrac{e^{-10} \times 10^3}{3!} = 0.0078$

3. The yield (in kg) per plot of a crop grown in 10,000 plots of a farm has mean and standard deviation of 40 kg and 5 kg respectively. If the yield of the plots is assumed to follow normal distribution, then find the expected no. of plots having yield (i) less than 30 kg (ii) more than 45 kg (iii) less than 50 kg (iv) between 35 and 50 kg.

Solution

Let the random variable X denotes the yield (in kg per plot). X ~ N(40 kg, 5 kg)

Then $Z = \dfrac{X - 40}{5} \sim N(0,1)$.

(i) When $X = 30$, $Z = \dfrac{30 - 40}{5} = -2$,

So, P(yield of plot less than 30 kg) $= P(X < 30) = P(Z < -2)$

$= P(Z < 0) - P(-2 < Z < 0) = 0.5 - \frac{1}{2}P(-2 < Z < 2)$ (due to symmetry of normal curve)

$= 0.5 - \dfrac{0.9544}{2} = 0.5 - 0.4772 = 0.0228$.

Thus the expected number of plots having yield less than 30 kg = 10,000 × $P(X < 30) = 10{,}000 \times 0.0228 = 228$

(ii) When $X = 45$, $Z = \dfrac{45 - 40}{5} = 1$;

So, P(yield of plot more than 45 kg) $= P(X > 45) = P(Z > 1)$

$= P(Z > 0) - P(0 < Z < 1) = 0.5 - \dfrac{1}{2}P(-1 < Z < 1)$ (due to symmetry of the curve) $= 0.5 - \dfrac{1}{2} \times 0.6826 = 0.1587$.

Thus the expected number of plots having yield more than 45kg = 10000 $P(X > 45) = 10000 \times 0.1587 = 1587$.

(iii) When $X = 50$, $Z = \frac{50-40}{2} = 2$

So, P (yield of plot less than 50 kg) = $P(X < 50) = P(Z < 2)$

$= P(Z < 0) + P(0 < Z < 2) = 0.5 + \frac{1}{2}P(-2 < Z < 2)$

(due to symmetry of normal curve)

$= 0.5 + \frac{0.9544}{2} = 0.5 + 0.4772 = 0.9772$

Thus the expected number of plots having yield less than 50 kg = 10000 × $P(X < 50)$ = 10000 × 0.9772 = 9772

(iv) When $X = 35$, $Z = \frac{35-40}{2} = -1$; When $X = 50$, $Z = \frac{50-40}{2} = 2$

So, P(yield of plot between 35 and 50 kg) = $P(35 < X < 50)$

$= P(-1 < Z < 2)$

$= P(-1 < Z < 0) + P(0 < Z < 2) = \frac{1}{2} \times P(-1<Z<1) + \frac{1}{2}P(-2 < Z < 2)$ (due to symmetry of normal curve)

$$= \frac{0.6826}{2} + \frac{0.9544}{2} = 0.3413 + 0.4772 = 0.8185.$$

Thus the expected no. of plots having yield between 35 and 50 kg = 10,000 × $P(35 < X < 50)$ = 10000 × 0.8185 = 8185.

Chapter 8

Correlation and Regression

Concept of correlation was given by Karl Pearson in 1890.

The study of strength and nature of relationship between two variables is called correlation. Two variables are said to be correlated, if the change in one variable affects the other variable. Correlation is also known as covariation.

Bivariate Distribution: Distribution involving two variables is known as bivariate distribution. For example, Height and weight of persons, yield and production of a crop in.

On basis of type of association between variables, correlation is of two types:

1. **Positive Correlation:** If the two variables deviate in the same direction, i.e., the increase (or decrease) in one variable results in a corresponding increase (or decrease) in other, correlation is said to be positive. e.g., Height and Weight of a person, Income and Expenditure of several families.
2. **Negative Correlation:** If the two variables deviate in the opposite direction i.e., the increase (or decrease) in one variable results in a corresponding decrease (or increase) in other, correlation is said to be negative.

   Example, Number of fruits & fruit size; degree of pest incidence and yield; etc.

Uncorrelated variables: If the change in one variable does not affect and is also not affected by change in the other variable, then the two variables are said to be uncorrelated e.g., Height of students and marks obtained by students are two uncorrelated variables.

Scatter diagram: The diagram of dots obtained by plotting the values of variables X & Y along X-axis & Y-axis are called scatter diagram. It is the simplest way of diagrammatic representation of bivariate data

Use of scatter diagram: From the scatter diagram, we can form a rough idea regarding the correlation between two variables. If the points are wider apart it indicates low degree of correlation and if the points are dense it indicates high degree of correlation.

Karl Pearson's correlation coefficient: It is a numerical measure of linear relationship between two variables X & Y and is defined as the ratio of covariance between the variables to the product of their standard deviations.

$$r(X.Y)=\frac{COV(X,Y)}{\sigma_X\sigma_Y}=\frac{\frac{1}{n}\sum_{i=1}^{n}(x_i-\overline{x})(y_i-\overline{y})}{\left[\frac{1}{n}\sum_{i=1}^{n}(x_i-\overline{x})^2\times\frac{1}{n}\sum_{i=1}^{n}(y_i-\overline{y})^2\right]^{1/2}}$$

On simplifying the above formula, we will get the short cut formula for $r(X, Y)$,

$$r(X,Y)=\frac{n\sum_{i=1}^{n}x_iy_i-\sum_{i=1}^{n}x_i\sum_{i=1}^{n}y_i}{\left[\left\{n\sum_{i=1}^{n}x_i^2-\left(\sum_{i=1}^{n}x_i\right)^2\right\}\left\{n\sum_{i=1}^{n}y_i^2-\left(\sum_{i=1}^{n}y_i\right)^2\right\}\right]^{1/2}}$$

$r(X, Y)$ can also be written as r_{XY}.

Karl Pearson's correlation coefficient is also known as linear correlation coefficient because it measures only the linear relationship between two variables. Karl Pearson's correlation coefficient is also known as simple correlation coefficient or product moment correlation coefficient.

Properties of Correlation coefficient

1. It measures the extent of linear relationship between the variables.
2. It measures the nature of relationship between the variables.
3. Its value ranges from –1 to +1.
4. For independent variables, correlation coefficient, $r = 0$.
5. It is independent of change of origin & scale (or units).
6. It is suitable for further statistical analysis. If the correlation coefficient between two variables is known for two or more classes, then these values can be used for finding the correlation coefficient between two variables for the combined classes.

   Combined correlation coefficient is given by,

$$r=\frac{\sum_{i=1}^{K}n_ir_i\sigma_{x_i}\sigma_{y_i}+\sum_{j=1}^{K}n_idx_idy_i}{n\sigma_x\sigma_y}\quad K\rightarrow\text{no. of classes}$$

Assumptions underlying Karl Pearson's correlation coefficient

1. The variables under study are linearly related.
2. Each of the variables must follow normal distribution.
3. Each variable must be affected by common forces. For example, height and weight are controlled by common force i.e., body metabolism. But height and intelligence are not controlled by common force.

Independent variables are uncorrelated but the converse may not be always be true. i.e., uncorrelated variable may not always be independent. This is because if $r = 0$, then it indicates the absence of linear relationship between the variables but there may be some other relationship between the variables such as quadratic ($y = x^2$), cubic ($y = x^3$), log arithmetic ($y = \log x$), etc.

Spurious or non-sense correlation: Sometimes, even if the forces operating on the two variable series are entirely independent of each other, the value of correlation coefficient for a given set of data of the two variables is not zero. Such correlation is termed as chance or spurious or non-sense correlation. For example, correlation between height of mango trees and height of the mango growers.

Interpretation of correlation coefficient

If $r > 0$, it indicates positive correlation.

If $r < 0$, it indicates negative correlation.

If $r = 0$, it indicates no correlation.

If $r = +1$, it indicates Perfect positive correlation.

If $r = -1$, it indicates Perfect negative correlation.

If r is close to 0, it indicates low degree of correlation.

If r is close to 1 or -1, it indicates high degree of correlation.

r_{XY} provides a measure of linear relationship between the variables X & Y.

Important interpretation of correlation coefficient is given by r^2, which is called coefficient of determination.

Utility of r^2: r^2 shows the amount of variability in one variable which can be explained by the variability in the second variable.

For example, if $r = 0.8$, then $r^2 = 0.64$ or 64%

It means that 64% of the variability in one variable can be explained by the variation in the other variable. Remaining 36% variation in the first variable is unexplained.

Probable Error – It is an old measure to test the reliability of correlation coefficient. $P.E.(r) = 0.6745 \times \frac{1-r^2}{\sqrt{n}}$

n = Number of observations; r = correlation coefficient.

Interpretation of P.E.(r)

1. If $r < PE(r)$, then r is considered insignificant.
2. If $r > 6PE(r)$, then r is considered definitely as significant.
3. If $PE(r) < r < 6PE(r)$, then no definite significance of r.

Thus PE(r) is used to test the significance of r.

Uses of correlation coefficient:

1. It determines the strength of linear relationship between the two variables.
2. It gives the nature of relationship between two variables – whether positive or negative.
3. It helps to determine the regression coefficients.
4. Square of the correlation coefficient (i.e., r^2) shows the amount of variation in one variable which can be explained by other variable.

Regression

The term regression was coined by Sir Francis Galton. The term regression means "stepping back towards average".

Definition: Regression analysis is a mathematical measure of the average relationship between two or more variables in terms of the original units of the data. In regression analysis there are two types of variables:

1. **Independent variable:** The variable which influences the values of other variable and is used for prediction of that variable is called independent variable. It is also known as predictor or regressor or explanatory variable.
2. **Dependent variable:** The variable whose value is influenced by the other variable and is predicted by that variable is called dependent variable. It is also known as predicted or regressed or explained variable.

For example, Rate of fertilizer affects yield of a crop. Thus rate of fertilizer is independent variable and yield is the dependent variable.

When the values of independent and dependent variables are given, we can find estimated or predicted values of the dependent variable by using the values of independent variables with the help of regression equation involving both the variables. Thus the dependent variables has two types of values:

(i) Given/actual/observed values

(ii) Predicted/estimated values: These are the values which are obtained by putting the values of the independent variable in the regression equation which involves both independent and dependent variables.

The regression equation involving the two variables (independent and dependent) can be represented in form of an equation.

$Y = a + b_{YX}X$; This is regression equation of Y on X. Here, Y is the dependent variable and X is the independent variable.

$X = a + b_{XY}\,Y$; This is regression equation of X on Y. Here X is the dependent variable and Y is the independent variable.

These regression equations are simple linear regression equations. Since the regression equations are linear so they can be represented in form of straight lines called lines of regression.

More specifically for regression equation of Y on X, slope b is denoted as b_{YX} and for regression equation of X on Y, slope b is denoted as b_{XY}.

There are two lines of regression: The regression equation of y on x is $Y = a + b_{YX}.X$ is established by minimizing the S.S.E. in y, not in x. So this equation gives the best estimate of Y for a specific value of X. But this equation, though it can also give an estimate of the random variable X for a specific value of Y, the estimate of X will not be the best estimate. Now to find the best estimate of the random variable X for a specific value of the random variable Y, we use the regression equation of X on Y i.e., $X = a + b_{XY}.Y$ which is established by minimizing the S.S.E. in X. So there are two lines of regression.

Regression coefficient: The change in the value of dependent variable corresponding to a unit change in the value of independent variable is called regression coefficient of the dependent variable on the independent variable.

It is of two types: b_{YX} (the regression coefficient of Y on X) and b_{XY} (the regression coefficient of X on Y)

$$b_{YX} = \frac{COV(X,Y)}{\sigma_X^2} = \frac{\frac{1}{n}\sum_{i=1}^{n}(x_i - \bar{x})(y_i - \bar{y})}{\frac{1}{n}\sum_{i=1}^{n}(x_i - \bar{x})^2}$$

$$b_{XY} = \frac{COV(X,Y)}{\sigma_Y^2} \frac{\frac{1}{n}\sum_{i=1}^{n}(x_i - \bar{x})(y_i - \bar{y})}{\frac{1}{n}\sum_{i=1}^{n}(y_i - \bar{y})^2}$$

b_{YX}, the regression coefficient of Y on X is defined as the change in the value of dependent variable (Y) corresponding to a unit change in the value of independent variable (X).

b_{XY}, the regression coefficient of X on Y is defined as the change in the value of dependent variable (X). Corresponding to a unit change in the value of independent variable (Y).

Another form of regression equation

If (i) X on $Y \rightarrow X - \bar{x} = b_{XY}(Y - \bar{y}) \Rightarrow X = \bar{x} + b_{XY}(Y - \bar{y})$

(ii) Y on $X \rightarrow Y - \bar{y} = b_{YX}(X - \bar{x}) \Rightarrow y = \bar{y} + b_{YX}(X - \bar{x})$

Relation between correlation coefficient and regression coefficient

$$b_{YX} = \frac{COV(X,Y)}{\sigma_X^2} = \frac{COV(X,Y)}{\sigma_X \sigma_Y} \times \frac{\sigma_Y}{\sigma_X} = r_{XY}\frac{\sigma_Y}{\sigma_X}$$

$$b_{XY} = \frac{COV(X,Y)}{\sigma_Y^2} = \frac{COV(X,Y)}{\sigma_X \sigma_Y} \times \frac{\sigma_X}{\sigma_Y} = r_{XY}\frac{\sigma_X}{\sigma_Y}$$

Properties of regression coefficient

1. Geometric Mean between the regression coefficients gives the correlation coefficients i.e., $r_{XY} = \sqrt{b_{XY} b_{YX}}$. This is called fundamental property.
2. The sign of b_{XY}, $b_{YX,}$ and r_{XY} are same. This property is called signature property.
3. If one of the regression coefficients is greater than unity, the other must be less than unity. If $b_{XY} > 1$, then $b_{YX} < 1$. This is called magnitude property.
4. A.M. of regression coefficient is greater than or equal to correlation coefficient. i.e., $\frac{1}{2}(b_{XY} + b_{YX}) > r$. This property is called Mean property.
5. Regression coefficient is independent of change of origin but not of scale.
6. If $U = \frac{X-a}{h}$ and $V = \frac{Y-b}{k}$, then $b_{XY} = \frac{h}{k} b_{UV}$; $b_{YX} = \frac{k}{h} b_{VU}$

Regression coefficient measures the change in the dependent variable per unit change in the independent variable.

Relationship between the two lines of regression with correlation coefficient.

1. If 'r' is close to 0, i.e., for low degree of correlation, the line of regression are wider apart.
2. If 'r' is close to ± 1 i.e., for high degree of correlation, the lines of regression are closer.
3. If $r = 0$, i.e., for uncorrelated variables, the lines of regression are perpendicular to each other.
4. If $r = \pm 1$, i.e., for perfectly correlated variables, the lines of regression coincide.

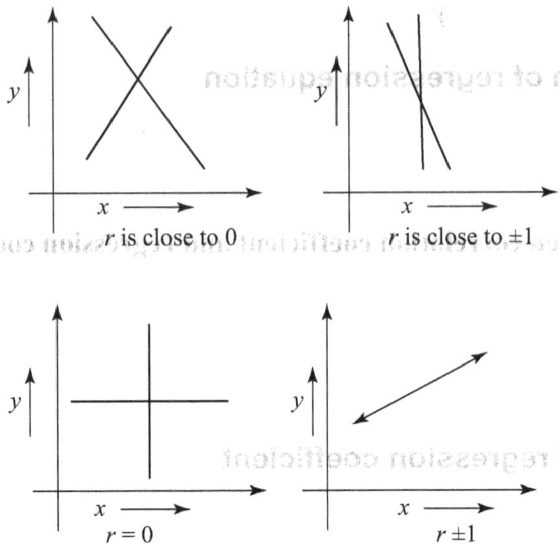

Low degree of correlation High degree of correlation Uncorrelated Perfect correlation.

Bivariate Frequency Distribution: The frequency distribution involving two variables is known as bivariate frequency distribution.

Correlation between X and Y is given by $r_{XY} = \dfrac{\text{cov}(X,Y)}{\sigma_X \sigma_Y}$

Regression coefficient of Y on X is given by $b_{YX} = \dfrac{\text{cov}(X,Y)}{\sigma_X^2}$ that for X on Y is given by $b_{XY} = \dfrac{\text{cov}(X,Y)}{\sigma_Y^2}$.

Difference Between Correlation and Regression

S. No.	Correlation	Regression
1.	It provides a measure of strength and nature of relationship between two variables.	It provides a measure of change in dependent variables per unit change in independent variable.
2.	No need to distinguish between independent and dependent variables i.e., r_{XY} is same as r_{YX}. So there is only one correlation coefficient between two variables.	Independent & dependent variables need to be determined separately i.e., b_{XY} is not same as b_{YX}. So there are two regression coefficients between two variables.
3.	Correlation coefficient is independent of both change of origin and scale.	Regression coefficient is independent of change of origin but not of scale.

Numerical Problems

1. If $b_{YX} = 0.2$; $b_{XY} = 3.2$; Find r_{XY}

   Solution: $r_{XY} = \sqrt{b_{XY} . b_{YX}} = \sqrt{0.2 \times 3.2} = 0.8$

2. If one of the regression coefficients is –3.2, find range of the other regression coefficient.

   Solution: As, $0 \le r_{XY} \le 1$; As $r_{XY} = b_{XY} \times b_{YX;}$ So, $0 \le b_{XY} . b_{YX} \le 1$; Since one of the regression coefficient is –3.2, the range of other regression coefficients is $-\frac{1}{3.2}$ to 0.

3. If one of the regression coefficients is 2, find range of the other regression coefficient.

   Solution :As, $0 \le r_{XY} \le 1$; As $r_{XY} = b_{XY.} b_{YX;}$ So, $0 \le b_{XY.} b_{YX} \le 1$; Since one of the regression coefficient is –2, the range of other regression coefficients is 0 to $\frac{1}{2}$.

4. If each value of X-series is divided by 10 and that of Y-series are divided by 5, how the regression coefficient is affected.

   Solution: $U = \frac{X}{10}; V = \frac{Y}{5}; b_{UV} = \frac{5}{10} \times b_{XY} = \frac{b_{XY}}{2}; b_{VU} = \frac{10}{5} \times b_{YX} = 2 \times b_{YX}$.

   Thus the new regression coefficient of X on Y (i.e., b_{UV}) is half of the original regression coefficient (b_{XY}) and that of Y on X (i.e., b_{VU}) is doubled.

5. Find most likely price in Mumbai corresponding to the price of Rs. 70 at Kolkata from the given data.

	Kolkata (X)	Mumbai (Y)
Average Price	$\bar{X}$ = Rs. 65.00	$\bar{Y}$ = Rs. 67.00
Standard Deviation	σ_X = Rs. 2.5	σ_Y = Rs. 3.5

$r_{XY} = 0.8$.

Solution: To estimate Y for given X, the regression equation Y on X needs to be established which is $(Y - \bar{Y}) = b_{YX}(X - \bar{X})$ i.e., $(Y - 67) = b_{YX}(X - 65)$

$$b_{YX} = \frac{\sigma_Y}{\sigma_X} r_{XY} = \frac{3.5}{2.5} \times 0.8 = 1.12$$

Thus the regression equation of Y on X is $(Y - 67) = 1.12(X - 65)$

i.e., $Y = -5.8 + 1.12\,X$

Putting $X = 70$ in the equation, we get $Y = -5.8 + 1.12 \times 70 = 72.6$

Thus the most likely price in Mumbai corresponding to the price of Rs. 70 at Kolkatta is Rs. 72.60.

6. The data given below provides the price and demand of 9 items.

Item no.	1	2	3	4	5	6	7	8	9
Price (in Rs.) (X)	200	250	280	360	240	280	320	400	440
Demand (in kg) (Y)	12	15	16	20	18	12	16	18	12

Fit the lines of regression of X on Y and Y on X. Estimate the price when demand = 25 kg. Also estimate the demand when price = Rs. 300.

Solution:

Item no.	**1**	**2**	**3**	**4**	**5**	**6**	**7**	**8**	**9**	
x_i	200	250	280	360	240	280	320	400	440	$\sum_{i=1}^{n} x_i = 2770$
y_i	12	15	16	20	18	12	16	18	12	$\sum_{i=1}^{n} y_i = 139$
$x_i y_i$	2400	3750	4480	7200	4320	3360	5120	7200	5280	$\sum_{i=1}^{n} x_i y_i = 43110$
x_i^2	40000	62500	78400	129600	57600	78400	102400	160000	193600	$\sum_{i=1}^{n} x_i^2 = 902500$
y_i^2	144	225	256	400	324	144	256	324	144	$\sum_{i=1}^{n} y_i^2 = 2217$

$$\overline{X} = \frac{\sum_{i=1}^{n} x_i}{n} = \frac{2770}{9} = 307.77; \overline{Y} = \frac{\sum_{i=1}^{n} y_i}{n} = \frac{139}{9} = 15.44$$

To estimate the price (X) for a given demand (Y) we have to fit the regression equation of X on Y.

Regression coefficient of X on Y,

$$b_{XY} = \frac{n\sum_{i=1}^{n} x_i y_i - \sum_{i=1}^{n} x_i \sum_{i=1}^{n} y_i}{n\sum_{i=1}^{n} y_i^2 - \left(\sum_{i=1}^{n} y_i\right)^2}$$

$$= \frac{9 \times 43110 - 2770 \times 139}{9 \times 2217 - (139)^2} = \frac{2960}{632} = 4.683$$

The regression equation of X on Y: $(X - \overline{X}) = b_{XY}(Y - \overline{Y})$

i.e., $(X - 307.77) = 4.683 \times (Y - 15.44)$

$\Rightarrow X = 307.77 - 4.683 \times 15.44 + 4.683 \times Y \Rightarrow X = 235.465 + 4.683 \times Y$

To estimate price for demand = 25 kg, we put Y =25 in the equation X = 235.465 + 4.683 × Y.

Thus, X = 235.465 + 4.683 × 25 = 235.465 + 117.075 = 352.54

Thus the estimated price for 25 kg demand is Rs. 352.54 i.e., Rs. 353 (approx.)

To estimate the demand (Y) for a given price (X) we have to fit the regression equation of Y on X.

Regression coefficient of Y on X, b_{YX} is obtained as

$$b_{YX} = \frac{n\sum_{i=1}^{n} x_i y_i - \sum_{i=1}^{n} x_i \sum_{i=1}^{n} y_i}{n\sum_{i=1}^{n} x_i^2 - \left(\sum_{i=1}^{n} x_i\right)^2} = \frac{9 \times 43110 - 2770 \times 139}{9 \times 902500 - (2770)^2}$$

$$= \frac{2960}{449600} = 0.0066$$

The regression equation of Y on $X (Y - \bar{Y}) = b_{YX} \times (X - \bar{X})$ i.e., (Y – 15.44) = 0.0066 × (X – 307.77) is obtained as

$\Rightarrow Y = 15.44 - 0.0066 \times 307.77 + 0.0066 \times X \Rightarrow Y = 13.409 + 0.0066 \times X$

To estimate the demand for price = Rs. 300, we put X = 300 in the equation Y = 13.409 + 0.0066 × X

Thus, Y = 13.409 + 0.0066 × 300 = 13.409 + 1.98 = 15.389

Thus the estimated demand for price of Rs. 300 is 15.389 kg.

Chapter 9

Introduction to Sampling and Test of Significance

The aggregate of data forming a subject of investigation is known as population. Example, All the crop plants grown in a plot; all the fruit plants in an orchard, etc. The population may be:

(i) **Finite:** No. of observations are definite. e.g., number of farms in a village, no. of house in a locality, no. of trees in an orchard are all finite population.

(ii) **Infinite:** Number of observation are indefinite e.g., (a) Number of insects in a region, number of grasses in a field etc. are all infinite population.

The Population may be:

(i) hypothetical or theoretical

Hypothetical population consists of unit that has no physical existence. This population exists only in imagination. For example, Result of tossing coin, throwing die,

(ii) Existent or real population consists of units that are physically present. This population actually exists. For example, Number of employees in a factory, Number of farms in a village.

In many instances, it is very difficult to study the entire population due to several constraints, such as, limited time, limited manpower, limited availability of resources, etc. In such cases we go for sampling. Sampling involves studying a portion of the population instead of studying the entire population and then the inferences is drawn for the population basing on the result obtained from the study of the sample. But the sample must be a representative one.

Sample: A portion of the population selected for the study to draw the conclusion regarding the entire population that is to estimate the characteristics of the population.

Parameter: Statistical measures based on all units of the population such as population mean (μ), Population standard deviation (σ), Population proportion (P), etc.

Statistic: Statistical measures based on the units included in the sample, such as sample mean $(\overline{x})$, Sample standard deviation(s), Sample proportion (p).

Sampling: The process of selection of the required no of units from a population for study and draw inferences regarding the population i.e., to estimate the population characteristics with a desired degree of precision.

Sample survey: The study of the unit included in the sample is called sample survey.

Sampled unit: List of all the population units from which sample is drawn is called as sampled unit.

Statistic provides an estimate of the parameter.

A Statistic is said to be an unbiased estimate of parameter if E (Statistic) = Parameter.

This means the mean of the statistic values obtained for all the possible samples of size, *n* that can be drawn from the population of size '*N*', is equal to the population parameter.

Need for sampling- Sampling is a necessary under the following conditions.

1. When population is infinite.
2. When population is hypothetical.
3. When testing is destructive.
4. When the results are required in short time.
5. When resources for the survey are limited particularly in respect of money and trained persons.

- Number of units included in the sample is sample size usually denoted by ***n***.
- Number of units in the population is population size usually denoted by *N*.

Three Types of Sampling

1. Random sampling

In this method of sampling, each unit of the population has same pre-assigned probability of being selected in the sample. It is known as probability sampling. The probability of selection of each unit may be equal or may be proportional to the sample size. Random sampling gives a representative sample of the population. The principle of statistics can be applied to random sampling.

2. Judgement or purposive sampling

If the samples are selected with purpose in view, it is known as judgement or purpose or subjective sampling. This method is subjected to biasness on the part of the investigation and does not give a representative sample of the population. So it is seldom used. This sampling procedure can be used if the user has proper and adequate knowledge on the subject matter. The sampled unit should be selected without any bias and with applicable justification.

3. Sampling distribution of the statistic

Number of samples of size n that can be drawn without replacement from a population of size '*N*' is *C*(*N*, *n*) or ${}^{N}C_{n}$. The arrangement of various values of a statistic obtained from different samples that can be drawn from a given population, along with their respective frequencies is called sampling distribution of the statistic. Let *C*(*N*, *n*) = *k*. Thus *k* different samples each of size '*n*' can be drawn from a population of size '*N*'.

Sampling distribution of '*k*' number of sample means is given as:

Sample mean $(\overline{x}_i)$	$\overline{x}_1$	$\overline{x}_2$	...	$\overline{x}_k$
No. of samples (f_i)	f_1	f_2	...	f_k

Sampling distribution of '*k*' numbers of sample variances is given as:

Sample variance s_i^2	s_1^2	s_2^2	...	s_k^2
No. of samples (f_i)	f_1	f_2	...	f_k

Standard error of a statistic

It is the standard deviation of the values of the statistic in the sampling distribution of the statistic.

Standard deviation of the values of mean obtained from different samples drawn from a population is standard error of sample mean.

This means, S.E. $(\overline{x}) = \sqrt{\text{Var}(\overline{x}_i)}$

But the formula for S.E.(sample mean) to be used practically is as follows:

If $\overline{x}$ represent the sample mean, then $SE(\overline{x}) = \frac{\sigma}{\sqrt{n}}; \sigma \rightarrow$ Population standard deviation, $\sigma = +\sqrt{\frac{1}{N}\sum_{i=1}^{n} f_i(x_i - \overline{x})^2}$ where $\overline{x}$ is the population mean, x_i is the value of ith unit in the population. But, in practice, σ is not known, so we use an estimate of σ which is provided by the Sample S.D. (s) to find the SE$(\overline{x})$. In that case the estimated value of SE$(\overline{x})$ is given by Est. SE$(\overline{x}) = \frac{s}{\sqrt{n}}$. For small sample ($n < 30$), the formula for Sample S.D. is given by, $s = \sqrt{\left(\frac{1}{n-1}\right)\sum_{i-1}^{n}(x_i - \overline{x})^2}$

Utility of Standard Error

1. It provides a limit for the expected variation in the population parameter (i.e., within which the population parameter is expected to lie) with a certain degree of precision.
2. It provides the precision or accuracy of the estimate.
3. Precision of estimate $\pm \frac{1}{\text{SE(estimate)}}$
4. It provides a value which could be compared with difference between the observed and expected value of the statistic, so as to declare the difference, significant or not, at a particular level of significance.

1. By using random number table select two simple random samples of 10 varieties (i) with replacement and (ii) without replacement from the population of 40 different varieties in which number of tillers per hill are given as below:

Variety No.	1	2	3	4	5	6	7	8	9	10	11	12	13	14	15	16	17	18	19	20
No. of tillers per hill	10	12	22	12	21	11	9	13	16	13	18	24	26	22	21	20	8	18	11	16
Variety No.	21	22	23	24	25	26	27	28	29	30	31	32	33	34	35	36	37	38	39	40
No. of tillers per hill	14	12	15	16	11	18	23	25	28	21	22	24	24	21	15	16	14	13	16	22

Solution:

The sampling units i.e., varieties are already numbered from 1 to 40. Population size, $N = 40$

Sample size, $n = 10$

Since $N = 40$ which is a two-digit number, a two-digit random number table is used to select the random sample. In direct approach the two-digit numbers from 01 to 40 are considered and the numbers greater than 40 are rejected. In remainder approach, the numbers greater than 40 in the random number table are divided by N (i.e., 40) and the remainder is considered. The selection of random numbers can be done in any way i.e., vertically downward or upward, horizontally to the right or to the left.

Direct approach			**Remainder approach**		
Random numbers taken from the table	**Selected random numbers**		**Random numbers taken from the table**	**Selected random numbers**	
	SRSWR	**SRSWOR**		**SRSWR**	**SRSWOR**
16	16	16	16	16	16
4	4	4	4	4	4
12	12	12	12	12	12
18	18	18	18	18	18

Direct approach			Remainder approach		
Random numbers taken from the table	Selected random numbers		Random numbers taken from the table	Selected random numbers	
	SRSWR	SRSWOR		SRSWR	SRSWOR
44	—	—	44	4	—
55	—	—	55	15	15
32	32	32	32	32	32
36	36	36	36	36	36
40	40	40	40	40	40
67	—	—	67	27	27
17	17	17	17		17
27	27	27	27		
17	17	—	17		
36		—	36		
32		—	32		
67		—	67		
24		24	24		

Test of Significance

The process of testing whether the difference between the observed sample statistic and expected population parameter or testing whether the difference between observed sample statistic of two or more samples are significant or not, at a specified level of significance.

- **Hypothesis**: A definite, doubtful and verifiable statement is called hypothesis.
- **Statistical hypothesis**: It is an assumption about a random variable or a set of random variables.
- **Null hypothesis**: It is a hypothesis of no difference and forms a basis for the test of significance. Denoted by H_0.

  Example: H_0: $\theta = \theta_0$ (for single sample); H_1: $\theta_1 = \theta_2$ (for two samples)

  θ is the actual population parameter (It is estimated from the sample)

  θ_0 is the assumed or expected population parameter.

  θ_1, θ_2 are population parameters estimated from two different samples
- **Alternative hypothesis**: - Any hypothesis which is complementary to null hypothesis. If null hypothesis is H_0: $\theta = \theta_0$, then alternative hypothesis will be any one of the following forms:

  $H_1: \theta \neq \theta_0$ (two-tailed test)

  $H_1: \theta > \theta_0$ (right-tailed test) } (one-tailed test)

  $H_1: \theta < \theta_0$ (left-tailed test)

  Sample standard deviation is given by, $s = +\sqrt{\frac{1}{n-1}\sum_{i=1}^{n}(x_i - \overline{x})^2}$;

x_i is the value of the ith unit in the sample; $\overline{x}$ - sample mean

For large samples ($n \geq 30$), sample S.D. is given by, $s = +\sqrt{\frac{1}{n}\sum_{i=1}^{n}(x_i - \overline{x})^2}$

Because in large samples, $n - 1 \approx n$.

If null hypothesis is H_0: $\theta_1 = \theta_2$, then alternative hypothesis will be of any one of the following forms:

H_1: $\theta_1 \neq \theta_2$ (two-tailed test)
H_1: $\theta_1 > \theta_2$(Right-tailed test) } (one-tailed test)
H_1: $\theta_1 < \theta_2$(left-tailed test) }

Thus alternate hypothesis enables us to decide the type of test, whether two-tailed or one-tailed (right-tailed or left tailed), to be used.

Errors in testing of hypothesis: Taking a wrong decision regarding the acceptance or rejection of $H_0$is known as error. It is of two types: -

Type-I error: Rejecting H_0 when it is true. Probability of Type I error is α. It is producer's risk.

Type-II error: Accepting H_0 when it is false. Probability of Type II error is β. It is consumer's risk.

Power of the test $= 1 - \beta$. Power of the test is the correct decision of rejecting the null hypothesis (H_0) when it is false.

Level of significance: It is the probability of committing type I error. It is generally fixed at 5% (i.e., 0.05) or 1% (i.e., 0.01). The fixation of level of significance depends on availability of experimental material, money, time, nature of problem dealt with, etc.

Type-II error is generally more serious than type-I error. Both the errors cannot be reduced simultaneously. We generally fix type-I error (which is less serious) at a certain controlled level and minimize the type-II error as far as possible.

Test statistic: The sampling distribution of a statistic such as Z, t, F, χ^2 which measure the extent to which the sample departs from H_0 in some relevant aspect are known as test statistic or test criterion.

The theoretical value of test statistic is called the critical value or tabulated value. The value of test statistic based on the sample data is called calculated value.

Degree of freedom (d.f.) –Degree of freedom is the number of observations that arc free to vary after certain restrictions have been placed on the data.

d.f. = number of obs. – number of restrictions imposed on the data

Confidence interval- It is the interval within which the population parameter is expected to lie with a certain level of confidence.

Confidence limits or fiducial limits – The lower and upper bounds of the interval estimates of a parameter are called confidence limits or fiducial limits.

For small samples,
(100 – α)% confidence or fiducial limits for population mean μ is given by,

$$\bar{x} - t_{\alpha,n-1} \times \frac{\sigma}{\sqrt{n}} \leq \mu \leq \bar{x} + t_{\alpha,n-1} \times \frac{\sigma}{\sqrt{n}}$$

For large samples,
(100 – α)% confidence or fiducial limits for population mean μ is given by,

$$\bar{x} - Z_{\alpha} \times \frac{\sigma}{\sqrt{n}} \leq \mu \leq \bar{x} + Z_{\alpha} \times \frac{\sigma}{\sqrt{n}}$$

Level of confidence: It is the probability that the parameter will lie within the assigned confidence interval.

Level of confidence = 1 – *P*(type (I) error) = 1 – level of significance = 1 – α.

Chapter 10

Small Sample Test

A sample is said to be small sample if the sample size(n) is less than 30. Various small sample tests are: (i) *t*-test (ii) *F*-test

t-test performs the following tests for small sample(s).

1. To test the equality between a single sample mean and the assumed population mean.
2. To test the equality of two independent sample means.
3. To test the equality of two paired sample means.
4. To test the equality between sample correlation coefficient and the assumed population correlation coefficient

Student's t-test: Student's *t*-test is used to test the equality between a single mean and the assumed population mean. This test is given by W.S. Gossett.

Here we are interested to know whether the population mean (μ) is equal to a specified value (μ_0).

The steps involved in the test are;

(*a*) Null hypothesis, H_0: $\mu = \mu_0$

(*b*) Alt. hypothesis, H_1: $\mu \neq \mu_0$ (two-tailed test)

H_1: $\mu > \mu_0$(right-tailed test)
H_1: $\mu < \mu_0$(left-tailed test) } (one-tailed test)

To test this hypothesis, a random sample of small size ($n < 30$) is drawn from the population. Sample mean ($\overline{x}$) is an estimate of μ and thus used to draw conclusion about μ.

Sample mean is given by, $\overline{x} = \frac{1}{n}\sum_{i=1}^{n} x_i$

(*c*) Level of significance (α) is specified generally at 5% (0.05) or 1% (0.01)

(*d*) Calculation of the test statistic:

Test statistic, $t = \frac{\overline{x} - \mu_0}{\sqrt{\frac{s^2}{n}}}$; s^2 is an unbiased estimate of σ^2

$s^2 \rightarrow$ Sample Variance; $\sigma^2 \rightarrow$ Population Variance. $s^2 = \frac{1}{n-1}\sum_{i=1}^{n}(x_i - \overline{x})^2$

(*e*) From t-table, the critical or tabulated value of t is obtained for $(n - 1)$ d.f at α level of significance. It is denoted as t_α

(*f*) **Conclusion:** If |Calculated $t| \geq t_\alpha$, then H_0 is rejected at α level of significance. Here t is said to be significant. Thus, sample mean $(\bar{x})$ differs significantly from the assumed population mean (μ_0).

Hence, it is finally concluded that there is significant difference between the actual population mean (μ) and assumed population mean (μ_0).

If |Calculated $t| < t_\alpha$, then H_0 is accepted at α level of significance. Here t is said to be non-significant. Thus, sample mean $(\bar{x})$ does not differ significantly from the assumed population mean (μ_0).

Hence, it is finally concluded that there is no significant difference between the actual population mean (μ) and assumed population mean (μ_0) and they can be treated as identical at α level of significance.

Assumptions for student's t-test

1. The parent population from which the sample is drawn is normal.
2. The sample observations are independent.
3. The population variance is unknown.

Problem: A farmer assumes that the average yield /plot of his farm is 50 kg. 15 plots are chosen at random from the farm and the yield of the plots are given as: 58, 46, 48, 60, 56, 55, 50, 42, 38, 66, 54, 56, 57, 60, 64. Test whether the mean yield of the plots is in accordance with the assumption of the farmer.

Solution

Let μ_0 be the assumed population mean. Given, $\mu_0 = 50$ kg. μ be the actual population mean.

An estimate of the actual population mean, μ is given by the sample mean $\bar{x}$.

Sample size, $n = 15$.

The steps of the test are:

(i) Null hypothesis, $H_0: \mu = \mu_0$

(ii) Alternative hypothesis, $H_1: \mu \neq \mu_0$ (two – tailed test)

(iii) Level of significance, $\alpha = 0.05$ (5%) or 0.01(1%)

(iv) Test statistics, $t = (\bar{x} - \mu_0)/SE(\bar{x})$

To find $\overline{x}$ and SE $(\overline{x})$

x_i	58	46	48	60	56	55	50	42	38	66	54	56	57	60	64	$\sum_{i=1}^{n} x_i = 810$
x_i^2	3364	2116	2304	3600	3136	3025	2500	1764	1444	4356	2916	3136	3249	3600	4096	$\sum_{i=1}^{n} x_i^2 = 44606$

Sample mean, $\overline{x} = \frac{\sum_{i=1}^{n} x_i}{n} = \frac{810}{15} = 54;$

Sample Variance, $s^2 = \frac{\sum_{i=1}^{n} x_i^2 - \frac{\left(\sum_{i=1}^{n} x_i\right)^2}{n}}{n-1} = \frac{44606 - \frac{(810)^2}{15}}{15-1} = 61.8571$

$$\text{S.E}(\overline{x}) = \sqrt{\frac{s^2}{n}} = \sqrt{\frac{61.8571}{15}} = 2.031.$$

Thus, $t = \frac{54-50}{2.031} = \frac{4}{2.031} = 1.969$

Thus calculated value of t, $t_{cal} = 1.969$; $|t_{CAl}| = 1.969$

Conclusion: Comparison of $|t_{cal}|$ with tabulated value of t at n - 1 = 15 − 1=14 d.f for 0.05 and 0.01 level of significance.

For two-tailed test, tabulated value of t at 14 d.f for 0.05 (5%) Level of significance, $t_{0.05} = 2.145$ tabulated value of t at 14 d.f for 0.01 (1%) Level of significance, $t_{0.01} = 2.977$

As, $|t_{cal}| < t_{0.05}$, the null hypothesis (H_0) is accepted at 5% level of significance. Here 't' is considered to be insignificant. This means the sample mean (i.e., estimated population mean) is not significantly different from the assumed population mean and the existing difference is due to chance fluctuation. Thus, it is concluded that mean yield of the plots is in accordance with the assumption of the farmer.

2. **Fisher's t – test is** used to test the equality of two independent sample means.

Here we are interested to know whether two population means μ_1 and μ_2 differ significantly or not. To test this, we take random samples of sizes n_1 and

n_2 from the two populations, where n_1 and n_2 are both less than 30 (i.e., we take small samples).

Let $x_{11}, x_{12}, \ldots, x_{1i}, \ldots, x_{1n1}$ be the observations of 1st sample.

$x_{21}, x_{22}, \ldots, x_{2i}, \ldots, x_{2n2}$ be the observations of 2nd sample

(a) Null hypothesis, $H_0: \mu_1 = \mu_2$ i.e., the two population means are identical.

(b) Alternative hypothesis, $H_1: \mu_1 \neq \mu_2$ (two tailed test)

$$\left.\begin{aligned} H_1 &: \mu_1 > \mu_2 \text{(right-tailed test)} \\ H_1 &: \mu_1 < \mu_2 \text{(left-tailed test)} \end{aligned}\right\} \text{One-tailed test}$$

we find the two sample means $\bar{x}_1$ and $\bar{x}_2$ to draw conclusions about μ_1 and μ_2.

Mean of 1st sample, $\bar{x}_1 = \frac{1}{n_1}\sum_{i=1}^{n_1} x_{1i}$; Mean of 2nd sample, $\bar{x}_2 = \frac{1}{n_2}\sum_{i=1}^{n_2} x_{2i}$

(c) Level of significance (α) is generally fixed at certain level (usually 5% or 1%).

(d) Test statistic, $t = \frac{\bar{x}_1 - \bar{x}_2}{SE(\bar{x}_1 - \bar{x}_2)}$; $\text{SE}(\bar{x}_1 - \bar{x}_2) = \sqrt{S^2\left(\frac{1}{n_1} + \frac{1}{n_2}\right)}$

Here S^2 is pooled variance and is given by:

$$S^2 = \frac{1}{n_1 + n_2 - 2}\left\{\sum_{i=1}^{n_1}(x_{1i} - \bar{x}_1)^2 + \sum_{i=1}^{n_2}(x_{2i} - \bar{x}_2)^2\right\}$$

(e) From the *t*-table, the critical value of t is obtained for $n_1 + n_2 - 2$ d.f. at α level of significance. It is denoted as t_α

(f) If $|\text{cal}t| \geq t_\alpha$ then '*t*' is considered significant. Here we reject H_0 and conclude that the sample means differ significantly. Hence, it is finally concluded that there is no significant difference between the two population mean (μ_1 and μ_2) If $|\text{cal}t| < t_\alpha$ then '*t*' is considered insignificant. Here we accept H_0 and conclude that sample means do not differ significantly and hence the two population means are identical.

Assumptions for Fisher's t-Test

(i) The parent population from which the samples are drawn is normal.

(ii) The sample observations are independent

(iii) The population variances are equal and unknown i.e., $\sigma_X^2 = \sigma_Y^2 = \sigma^2$, where σ^2 is unknown.

Problem

Two different varieties (*A* and *B*) of a fruit are grown. in two different farms. Test whether or not the mean yield/plant (in Quintals) of the two varieties differ significantly. The yield/plant (in Quintals) of the two varieties are given as:

Variety A	30	45	46	48	50	60	55	54	64	66	70	62
Variety B	18	24	44	56	66	54	43	29	40	44	-	-

Solution

Let the random variable *X* denote the yield/plant (in kg) of variety A and the random variable *Y* denote the yield/plant (in kg) of variety *B*.

Sample size of the random variable *X*, n_1 =12; Sample size of the random variable Y, n_2=10;

Let μ_X and μ_Y are the population mean yield of variety A and B respectively.

The steps involved in the test are:

(i) Null hypothesis, $H_0 : \mu_X = \mu_Y$

(ii) Alternative hypothesis, $H_1 : \mu_X \neq \mu_Y$ (two - tailed test)

(iii) Level of significance, $\alpha = 0.05$ (5%) and 0.01(1%) (iv) Test Statistic,

$$t = \frac{\bar{x}}{\text{SE}(\bar{x} - \bar{y})}$$

To find $\bar{x}, \bar{y}, \text{SE}(\bar{x} - \bar{y})$:

(x_i)	30	45	46	48	50	60	55	54	64	66	70	62	$\sum_{i=1}^{n_1} x_i = 650$
(x_i^2)	900	2025	2116	2304	2500	3600	3025	2916	4096	4356	4900	3844	$\sum_{i=1}^{n_1} x_i^2 = 36582$

(y_i)	18	24	44	56	66	54	43	29	40	44	$\sum_{i=1}^{n_2} y_i = 418$
(y_i^2)	324	576	1936	3136	4356	2916	1849	841	1600	1936	$\sum_{i=1}^{n_2} y_i^2 = 19470$

$$\bar{x} = \frac{\sum_{i=1}^{n_1} x_i}{n_1} = \frac{650}{12} = 54.17; \bar{y} = \frac{\sum_{i=1}^{n_2} y_i}{n_2} = \frac{418}{10} = 41.80$$

$$\text{SE}(\bar{x} - \bar{y}) = \sqrt{S^2\left(\frac{1}{n_1} + \frac{1}{n_2}\right)}$$; S^2 is the pooled variance;

$$S^2 = \frac{1}{n_1 + n_2 - 2}\left[\left\{\sum_{i=1}^{n_1} x_i^2 - \frac{\left(\sum_{i=1}^{n_2} x_i\right)^2}{n_1}\right\} + \left\{\sum_{i=1}^{n_2} y_i^2 - \frac{\left(\sum_{i=1}^{n_2} y_i\right)^2}{n_2}\right\}\right]$$

$$S^2 = \frac{1}{12+10-2}\left[\left\{36582 - \frac{(650)^2}{12}\right\} + \left\{19470 - \frac{(418)^2}{10}\right\}\right] = 168.5633;$$

$$\text{SE}(\bar{x} - \bar{y}) = \sqrt{168.5633\left(\frac{1}{12} + \frac{1}{10}\right)} = 5.559$$

$$t = \frac{\bar{x} - \bar{y}}{SE(\bar{x} - \bar{y})} = \frac{54.17 - 41.80}{5.56} = \frac{12.37}{5.559} = 2.225;$$

Thus, calculated value of t, $t_{cal} = 2.25$; $|t_{cal}| = 2.225$

Conclusion: Comparison of $|t_{cal}|$ with tabulated value of t at $n_1 + n_2 - 2 = 12 + 10 - 2 = 20$ d.f for 0.05 and 0.01 level of significance.

For two-tailed test, tabulated value of t at 20 d.f for 0.05 (5%) level of significance, $t_{0.05} = 2.086$; tabulated value of t at 20 d.f for 0.01 (1%) level of significance, $t_{0.01} = 2.845$.

As $t_{0.05} < |t_{cal}| < t_{0.01}$ null hypothesis (H_0) is rejected at 5 % level of significance. Here 't' is considered to be significant at 5% level of significance. This means the two sample means differ significantly at 5% level of significance. Thus, it is concluded that the yield of two varieties (A and B) of the fruit differ significantly at 5% level of significance.

Paired t-test

This test procedure is used for comparing the population means when the samples are not independent i.e., the samples are correlated. Here the sample sizes of both the samples are equal i.e., $n_1 = n_2 = n$. The data are taken in pairs.

The sample observations are paired together i.e., the pair of observations (x_i, y_i), $i = 1, 2, \ldots n$ corresponds to the same sample unit (i^{th} sample unit) for both the samples. Suppose, in feeding experiments, we want to test the efficiency of a particular feed in increasing the body weight of animals. Thus, we want to test whether mean weight of animals before the food (μ_X) and after the food (μ_Y) are same or not.

For this we select 'n' pairs of observations

Let x_i and y_i ($i = 1, 2, \ldots, n$) be the readings of observations corresponding to i^{th} unit, respectively.

(a) Null hypothesis, $H_0 : \mu_X = \mu_Y$ (i.e., the mean weight before and after the feed do not differ significantly)

(b) Alternative hypothesis, $H_1 : \mu_X \neq \mu_Y$ (Two- tailed test)

$H_1 : \mu_X > \mu_Y$ (Right-tailed test)
$H_1 : \mu_X < \mu_Y$ (Left-tailed test) } (one-tailed test)

(c) Level of significance (α) is generally fixed at certain level (usually 5% or 1%).

(d) Test statistic, $t = \dfrac{\bar{d}}{\sqrt{\dfrac{s_d^2}{n}}}$ where, $\bar{d} = \dfrac{1}{n}\sum_{i=1}^{n} d_i;;$

$d_i = x_i - y_i;\ d_i$ is the difference in values of the ith observation

$$s_d^2 = \frac{1}{n-1}\sum_{i=1}^{n}(d_i - \bar{d})^2$$

(e) From the t-table, the critical value of t is obtained for $n - 1$ d.f. at α level of significance. It is denoted as $t\alpha$

Conclusion: If $|t_{cal}| \geq t_\alpha$ then 't' is considered significant. Here we reject H_0 and conclude that the sample means differ significantly. Hence, it is finally concluded that there is no significant difference between the two population mean (μ_X and μ_Y).

If $|t_{cal}| < t_\alpha$ then 't' is considered insignificant. Here we accept H_0 and conclude that sample means do not differ significantly and hence the two population means are identical.

Problem

A particular diet was expected to increase the weight of cattle. 10 animals were selected from a population of cattle. The weights of selected animals were recorded. Then they are given the diet for one month. After one month weight of the same set of animals are recorded. The data are given below:

Cow No.	1	2	3	4	5	6	7	8	9	10
Weight before the diet (in kg)	680	667	690	730	750	766	760	780	700	740
Weight after the diet (in kg)	740	720	730	760	800	820	820	800	750	790

Test whether or not the diet increases the weight significantly.

Solution

Let the random variable X denotes the weight (in kg) of the animals before the diet was given and Y denotes the weight (in kg) of the animals after the diet was given. Number of observations, $n = 10$.

Let μ_X and μ_Y denote the mean weight of the population before and after the diet, respectively. $\overline{x}$ and $\overline{y}$ be the sample mean weight of the animals before and after the diet, respectively. The steps involved in the test are:

(i) Null hypothesis, $H_0: \mu_X = \mu_Y$

(ii) Alternative hypothesis, $H_0: \mu_X < \mu_Y$ (one-tailed test)

(iii) Level of significance, $\alpha = 0.05$ (5%) and 0.01 (1%);

(iv) To find the calculated value of the test statistic

Test statistic, $t = \dfrac{\overline{d}}{SE(\overline{d})}; \overline{d} = \dfrac{\sum_{i=1}^{n} d_i}{n}$; where, $d_i = x_i - y_i$

$$s_d^2 = \frac{1}{n-1}\left\{\sum_{i=1}^{n} d_i^2 - \frac{\left(\sum_{i=1}^{n} d_i\right)^2}{n}\right\}$$

Cow No.	1	2	3	4	5	6	7	8	9	10	
Weight before the diet (in kg) (x_i)	680	667	690	730	750	766	760	780	700	740	
Weight after the diet (in kg) (y_i)	700	710	723	740	770	820	800	800	720	750	
$d_i = x_i - y_i$	−20	−43	−33	−10	−20	−54	−40	−20	−20	−10	$\sum_{i=1}^{n} d_i = -270$
d_i^2	400	1849	1089	100	400	2916	1600	400	400	100	$\sum_{i=1}^{n} d_i^2 = 9254$

Thus, $$\overline{d} = \frac{-270}{10} = -27; s^2 = \frac{1}{10-1}\left\{9254 - \frac{(-270)^2}{10}\right\} = 218.22;$$

$$SE(\overline{d}) = \sqrt{\frac{218.22}{10}} = 4.671; t = \frac{-27}{4.671} = -5.78; |t| = 5.78$$

Thus, Calculated value of t, $t_{cal} = 5.78$

Conclusion: Comparison of t_{cal} with tabulated value of t at n-1 = 10 − 1 = 9 d.f. for 0.05 and 0.01 level of significance.

For one-tailed test,

tabulated value of t at 9 d.f for 0.05 (5%) Level of significance, $t_{0.05} = 1.833$

tabulated value of t at 9 d.f for 0.01 (1%) Level of significance, $t_{0.01} = 2.874$

As $|t_{cal}| > t_{0.01}$, null hypothesis (H_0) is rejected at 1 % level of significance. Here 't' is considered to be significant at 1 % level. This means the sample mean $\bar{x}$ is significantly less than the sample mean $\bar{y}$ at 1 % level of significance. Thus, it is concluded that the mean weight of the population (cows) after the diet has increased significantly as compared to the mean weight of the population (cows) before the diet, at 1% level of significance.

t-test for testing the significance of an observed sample correlation coefficient

Let r is the observed correlation in a sample of n pairs of observations.

ρ is the population correlation coefficient.

We want to test whether ρ differs significantly from 0 or not.

(a) Null hypothesis, $H_0 : \rho = 0$;

(b) Alternative hypothesis, $H_1 : \rho \neq 0$ (two tailed)

$H_1 : \rho > 0$ (right-tailed test)
$H_1 : \rho < 0$ (left-tailed test) } (one-tailed test)

Level of significance, $\alpha = 0.05$ (5%) or 0.01(1%), usually

Test Statistic, $t = \dfrac{r}{SE(r)}$; $\text{SE}(r) = \sqrt{\dfrac{1 - r^2}{n - 2}}$

From the t-table, the critical value of t is obtained for $n - 1$ d.f. at α level of significance. It is denoted as t_α

Conclusion: If $|t_{cal}| \geq t_\alpha$ then 't' is considered significant. Here we reject H_0 and conclude that the sample means differ significantly. Hence, it is finally concluded that there is no significant difference between the two population mean (μ_X and μ_Y).

If $|t_{cal}| < t_\alpha$ then 't' is considered insignificant. Here we accept H_0 and conclude that sample means do not differ significantly and hence the two population means are identical.

Problem

Find the correlation coefficient between no. of panicles per hill and the yield of paddy (in kg/ha) as per the data given below:

No. of Panicles/hill	25	22	34	33	45	43	23	47	43
Yield	1800	1600	2000	2100	3000	2900	1800	3500	3200

Also test the significance of correlation coefficient.

Solution

Let the random variable X denotes the number of panicles per hill and Y denotes the yield of paddy (in kg/ha).

Number of pairs of observations, n = 9

x_i	25	22	34	33	45	43	23	47	43	$\sum_{i=1}^{n} x_i = 315$
y_i	1800	1600	2000	2100	3000	2900	1800	3500	3200	$\sum_{i=1}^{n} y_i = 21900$
x_i^2	625	484	1156	1089	2025	1849	529	2209	1849	$\sum_{i=1}^{n} x_i^2 = 11815$
y_i^2	3240000	2560000	4000000	4410000	9000000	8410000	3240000	12250000	10240000	$\sum_{i=1}^{n} y_i^2 = 57350000$
$x_i y_i$	45000	35200	68000	69300	135000	124700	41400	164500	137600	$\sum_{i=1}^{n} x_i y_i = 820700$

Correlation Coefficient between X and Y,

$$r(X, Y) = \frac{n\sum_{i=1}^{n} x_i y_i - \sum_{i=1}^{n} x_i \sum_{i=1}^{n} y_i}{\left[\left\{n\sum_{i=1}^{n} x_i^2 - \left(\sum_{i=1}^{n} x_i\right)^2\right\}\left\{n\sum_{i=1}^{n} y_i^2 - \left(\sum_{i=1}^{n} y_i\right)^2\right\}\right]^{\frac{1}{2}}}$$

$$= \frac{9 \times 820700 - 315 \times 21900}{\left[\left\{9 \times 11815 - (315)^2\right\}\left\{9 \times 57350000 - (21900)^2\right\}\right]^{\frac{1}{2}}}$$

$$= \frac{487800}{509705.2} = 0.957$$

Test of significance of $r(X, Y)$: Let the population correlation coefficient be represented by ρ

Null hypothesis, H_0: $\rho = 0$; Alternative hypothesis, H_1: $\rho \neq 0$ (two-tailed test)

Level of significance, $\alpha = 0.05$ (5%) and 0.01 (1%)

Test statistic,

$$t = \frac{r}{SE(r)}; SE(r) = \sqrt{\frac{1-r^2}{n-2}} = \sqrt{\frac{1-0.957^2}{9-2}} = \sqrt{\frac{1-0.916}{7}} = \sqrt{\frac{0.084}{7}} = 0.109$$

$$\therefore t = \frac{0.957}{0.109} = 8.779$$

Conclusion: Calculated value of $t = 8.779$. Thus, $|t| = 8.779$

For two-tailed test: Tabulated value of t at 0.05 level of significance and 7 d.f, $t_{0.05} = 2.365$.

Tabulated value of t at 0.01(1%) level of significance and 9 – 2 = 7 d.f, $t_{0.01} = 3.499$.

As $|t| > t_{0.01}$, So H_0 is rejected at 0.01 (1%) level of significance. here 't' is considered to be significant at 1 % level. Thus it is concluded that the correlation coefficient between no. of panicles per hill and the yield of paddy is significant at 1 % level of significance.

F test

F test is used (i) to test the equality of two population variances (ii) to test the equality of more than two independent sample means.

In this chapter, we will study about use of F test for testing the equality of two population variances. Use of F test for testing the equality of two population variances is given by Snedecor. Hence, this test is known as Snedecor's *F* test.

Suppose we want to test whether two independent estimates of the population variance are homogeneous or not.

To test this, we take random samples of sizes n_1 and n_2 from the two populations, where n_1 and n_2 are both less than 30 (i.e., we take small samples).

Let $x_{11}, x_{12,\ldots}, x_{1i,\ldots} x_{1n_1}$ be the observations of 1st sample.

$x_{21}, x_{22,}\ldots, x_{2i,\ldots} x_{2n_2}$ be the observations of 2nd sample

The steps followed for the test are;

(a) Null hypothesis, $H_0 : \sigma_x^2 = \sigma_y^2$ i.e., the population variances are equal.

(b) Alternative hypothesis, H_1: $\sigma_X^2 \neq \sigma_Y^2$ (two-tailed test)

H_1: $\sigma_X^2 > \sigma_Y^2$ (right-tailed test)
H_1: $\sigma_X^2 < \sigma_Y^2$ (left-tailed test) } (one-tailed test)

(c) Level of significance (α) is generally fixed at certain level (usually 5% or 1%).

(d) Test statistic, $F = \frac{s_X^2}{s_Y^2}$, If $s_X^2 > s_Y^2$; $F = \frac{s_Y^2}{s_X^2}$, If $s_Y^2 > s_X^2$

Where, $s_X^2 = \frac{1}{n_1 - 1}\sum_{i=1}^{n_1}(x_i - \overline{x})^2$; s_X^2 - Sample Variance of X

$s_Y^2 = \frac{1}{n_2 - 1}\sum_{i=1}^{n_2}(y_i - \overline{y})^2$; s_Y^2 -Sample Variance of Y

From the *F*-table, the critical value of F is obtained for $n_1 - 1$ and $n_2 - 1$ d.f. at α level of significance. It is denoted as F_α

Conclusion: If $F_{cal} \geq F_\alpha$ then '*F*' is considered significant. Here we reject H_0 and conclude that the sample variances differ significantly. Hence, it is finally concluded that there is no significant difference between the two population variances (σ_X^2 and σ_Y^2).

If $|F_{cal} < F_\alpha$ then 'F_α' is considered insignificant. Here we accept H_0 and conclude that sample variances do not differ significantly and hence the two population variances are identical.

Problem:

Two different varieties (*A* and *B*) of a fruit are grown. in a farms. Test whether the variance of yield per plant (in Quintals) of the two varieties differ significantly or not. The yield per plant (in Quintals) of the two varieties is given as:

Variety A	40	55	66	68	80	70	65	64	74	76	80	72
Variety B	38	44	54	66	76	64	53	39	50	54	-	-

Solution: Let σ_X^2 and σ_Y^2 denotes the standard deviation of yield/plant of variety *A* and *B* respectively.

Here, $n_1 = 12$; $n_2 = 10$

Null hypothesis, H_0: $\sigma_X^2 = \sigma_Y^2$; Alternative hypothesis, H_1: $\sigma_X^2 \neq \sigma_Y^2$ (two-tailed test);

Level of significance, $\alpha = 0.05$ (5%) or 0.0 (1%)1;

To find the calculated value of the test statistic, *F*.

(x_i)	40	55	66	68	80	70	65	64	74	76	80	72	$\sum_{i=1}^{n_1} x_i = 810$
(x_i^2)	1600	3025	4356	4624	6400	4900	4225	4096	5476	5776	6400	5184	$\sum_{i=1}^{n_2} x_i^2 = 56062$
(y_i)	38	44	54	66	76	64	53	39	50	54		$\sum_{i=1}^{n_2} y_i = 538$	
(y_i^2)	1444	1936	2916	4356	5776	4096	2809	1521	2500	2916		$\sum_{i=1}^{n_2} y_i^2 = 30270$	

$$s_X^2 = \frac{1}{n_1 - 1}\left(\sum_{i=1}^{n_1} x_i^2 - \frac{\left(\sum_{i=1}^{n_1} x_i\right)^2}{n_1}\right) = \frac{1}{12-1}\left(56062 - \frac{810^2}{12}\right) = 126.091;$$

$$s_Y^2 = \frac{1}{n_2 - 1}\left(\sum_{i=1}^{n_2} y_i^2 - \frac{\left(\sum_{i=1}^{n_2} y_i\right)^2}{n_2}\right) = \frac{1}{10-1}\left(30270 - \frac{538^2}{10}\right) = 147.289$$

As $s_Y^2 > s_X^2$

$$\therefore\ F = \frac{s_Y^2}{s_X^2} = \frac{147.289}{126.091} = 1.168$$

Conclusion: Calculated $F = 1.168$.
In case of two-tailed test,
Tabulated value of F at 5 % (0.05) level of significance for 9,11 d.f, $F_{0.05} = 2.896$
Tabulated value of F at 1% (0.01) level of significance for 9,11 d.f, $F_{0.01} = 4.632$

As, Calculated $F < F_{0.05}$ for 9,11 d.f, we accept H_0 at 5 % level of significance. Thus F is considered to be insignificant and it is finally concluded that the two population variances do not differ significantly.

Chapter 11

Large Sample Test

Samples having size, $n \geq 30$ are known as large sample. For large values of n (i.e., for $n \geq 30$), almost all the distributions are very closely approximated by normal distribution. Thus in this case we apply the normal test, which is based upon the area property of the normal probability curve. Large Sample Test is also known as Z-test or SND (Standard Normal Deviate) test.

Z-test performs the following tests for large sample(s).

1. To test the equality between a single sample mean and the assumed population mean.
2. To test the equality of two independent sample means.

Test of significance for single mean: We are interested to know whether or not the population mean (μ) is equal to a specified value (μ_0).

To test this hypothesis, a random sample of large size ($n \geq 30$) is drawn from the population. Sample mean ($\bar{x}$) is an estimate of μ and thus used to draw conclusion about μ. Sample mean, $\bar{x} = \frac{1}{n}\sum_{i=1}^{n} x_i$. The steps involved in the test are:

The steps involved in the test are;

(a) Null hypothesis, H_0: $\mu = \mu_0$

(b) Alt. hypothesis, H_1: $\mu \neq \mu_0$ (two-tailed test)

H_1: $\mu > \mu_0$ (right-tailed test)
H_1: $\mu < \mu_0$ (left-tailed test) } (one-tailed test)

(c) Level of significance (α) is specified generally at 5% (0.05) or 1% (0.01)

(d) Calculation of the test statistic:

Test statistic, $t = \dfrac{\bar{x} - \mu_0}{\sqrt{\dfrac{s^2}{n}}}$; s^2 is an unbiased estimate of σ^2

$s^2 \rightarrow$ Sample Variance; $\sigma^2 \rightarrow$ Population Variance. $s^2 = \frac{1}{n}\sum_{i=1}^{n}(x_i - \bar{x})^2$

(e) From t-table, the critical or tabulated value of Z is obtained at α level of significance. It is denoted as Z_α

Conclusion: If Calculated $Z| \geq Z_\alpha$, then H_0 is rejected at α level of significance. Here Z is said to be significant. Thus Sample mean $(\bar{x})$ differs significantly from the assumed population mean (μ_0).Hence, it is finally concluded that there is significant difference between the actual population mean (μ) and assumed population mean (μ_0).

If |Calculated $Z| < Z_\alpha$, then H_0 is accepted at α level of significance. Here Z is said to be non-significant. Thus, Sample mean $(\bar{x})$ does not differ significantly from the assumed population mean (μ_0).

Hence, it is finally concluded that there is no significant difference between the actual population mean (μ) and assumed population mean (μ_0) and they can be treated as identical at α level of significance.

95% and 99 % confidence or fiducial limits for population mean μ is given by

$\bar{x} - 1.96(\sigma/\sqrt{n}) \leq \mu \leq \bar{x} + 1.96(\sigma/\sqrt{n})$ and $\bar{x} - 2.58(\sigma/\sqrt{n}) \leq \mu \leq \bar{x} + 2.58(\sigma/\sqrt{n})$, respectively.

Test of significance for difference of two means

Here we are interested to know whether two population means μ_1 and μ_2 are identical or not. To test it, we take random sample of sizes n_1 and n_2 from the two populations where n_1 and n_2 are both greater than or equal to 30 (i.e., we take large samples). The steps involved in the test are;

(a) Null hypothesis, H_0: $\mu_1 = \mu_2$

(b) Alternative hypothesis, H_1: $\mu_1 \neq \mu_2$ (two-tailed test)

H_1: $\mu_1 > \mu_2$ (right tailed test)
H_1: $\mu_1 < \mu_2$ (left tailed test) } (one-tailed test)

We find two sample means $\bar{x}_1$ and $\bar{x}_2$ to draw conclusions about μ_1 and μ_2, respectively.

(c) Level of significance, $\alpha = 0.05$ or 0.01

(d) Test statistic, $Z = \dfrac{\bar{x}_1 - \bar{x}_2}{\sqrt{\dfrac{\sigma_1^2}{n_1} + \dfrac{\sigma_2^2}{n_2}}}$.

If $\sigma_1^2 = \sigma_2^2 = \sigma^2$, then $\dfrac{\bar{x}_1 - \bar{x}_2}{\sigma\sqrt{\left(\dfrac{1}{n_1} + \dfrac{1}{n_2}\right)}}$

If σ_1^2 and σ_2^2 are unknown, then their estimates s_1^2 and s_2^2 are used.

Thus , in this case where σ_1^2 and σ_2^2 are unknown, $Z = \dfrac{\bar{x}_1 - \bar{x}_2}{\sqrt{\dfrac{s_1^2}{n_1} + \dfrac{s_2^2}{n_2}}}$

If σ_1^2 and σ_2^2 are unknown and equal, then an unbiased estimate of σ^2 is given by-

$$\hat{\sigma}^2 = \frac{n_1 s_1^2 + n_2 s_2^2}{n_1 + n_2} \quad \text{Thus, } Z = \frac{\bar{x}_1 - \bar{x}_2}{\hat{\sigma}\sqrt{\left(\dfrac{1}{n_1} + \dfrac{1}{n_2}\right)}}$$

(e) From the *t*-table, the critical value of *Z* is obtained. at α level of significance. It is denoted as Z_α

(f) If |cal *Z*| $\geq Z_\alpha$ then '*Z*' is considered significant. Here we reject H_0 and conclude that the sample means differ significantly. Hence, it is finally concluded that there is no significant difference between the two population mean (μ_1 and μ_2)

If If |cal *Z*| $\geq Z_\alpha$ then '*Z*' is considered insignificant. Here we accept H_0 and conclude that sample means do not differ significantly and hence the two population means are identical.

Problem: If the mean of a sample of size 40 is found to be 200. The sample variance is found to be 100. Test whether the population mean could be assumed to be 210 at 5% level and 1% 13v31 of significance.

Solution: Here, sample mean, $\bar{x} = 200$; $n = 40$; $s^2 = 100$.

Assumed population mean, μ_0

H_0: $\mu = 210$. H_1: $\mu \neq 210$

Level of significance, α = 5% and 1%;

Test statistic, $Z = \dfrac{200 - 210}{\sqrt{100/40}} = \dfrac{-10}{1.581} = -6.324; |Z| = |-6.324| = 6.324$

For two tailed test. tabulated value of *Z* at 5% level of significance, $Z_{0.05}$ = 1.96 tabulated value of *Z* at 1% level of significance, $Z_{0.01}$ = 2.58.

As, $|Z| > Z_{0.01}$, so H_0 is rejected at 1% level of significance and we conclude that the sample mean differs significantly from assumed population mean. Thus the population mean cannot be assumed to be 210.

Problem: Mean of two independent samples of sizes 35 and 40 are 120 and 150, respectively. Test whether the population means from which the samples are taken differ significantly or not, under the following conditions:

(i) Variances of the two populations are 160 and 360 respectively.

(ii) Variances of both the populations are identical and equal to 250.

(iii) Variances of the populations are unequal and unknown and their corresponding sample variances are 120 and 300, respectively.

(iv) Variances of the populations are assumed to be equal and unknown and their corresponding sample variances are 140 and 320, respectively

Solution: Given, $\bar{x}_1 = 120; \bar{x}_2 = 150;$

Null hypothesis, $H_0: \mu_1 = \mu_2$

Alternative hypothesis, $H_1: \mu_1 \neq \mu_2$ (two-tailed test)

Level of significance, $\alpha = 0.05$ and 0.01.

Computation of test statistic:

(i) $\sigma_1^2 = 16; \sigma_2^2 = 36; Z = \dfrac{\bar{x}_1 - \bar{x}_2}{\sqrt{\dfrac{\sigma_1^2}{n_1} + \dfrac{\sigma_2^2}{n_2}}} = \dfrac{120 - 150}{\sqrt{\dfrac{160}{35} + \dfrac{360}{40}}} = \dfrac{-30}{3.684} = -8.14$

$|Z| = 8.143$. For two-tailed test, $Z_{0.05} = 1.96$ and $Z_{0.01} = 2.58$

As, $|Z| > Z_{0.01}$, H_0 gets rejected thus the two sample means differ significantly and hence it is concluded that the two population means differ significantly.

(ii) $\sigma^2 = 250;\ Z = \dfrac{120 - 150}{\sqrt{250 \times (\dfrac{1}{35} + \dfrac{1}{40})}} = \dfrac{-30}{3.659} = -8.198$

$|Z| = 8.198$; For two-tailed test, $Z_{0.05} = 1.96$ and $Z_{0.01} = 2.58$

As, $|Z| > Z_{0.01}$, H_0 gets rejected thus the two sample means differ significantly and hence it is concluded that the two population means differ significantly.

(iii) $s_1^2 = 120;\ s_2^2 = 300\ \ Z = \dfrac{\bar{x}_1 - \bar{x}_2}{\sqrt{\dfrac{s_1^2}{n_1} + \dfrac{s_2^2}{n_2}}} = \dfrac{120 - 150}{\sqrt{\dfrac{120}{35} + \dfrac{300}{40}}} = \dfrac{-30}{3.306} = -9.075$

$|Z| = 9.075$; For two-tailed test, $Z_{0.05} = 1.96$ and $Z_{0.01} = 2.58$

As, $|Z| > Z_{0.01}$, H_0 gets rejected thus the two sample means differ significantly and hence it is concluded that the two population means differ significantly.

(iv) $s_1^2 = 140;\ s_2^2 = 320;$ since population variances are assumed to ** equal, we have to find pooled sample variance.

Pooled variance, $S^2 = \dfrac{n_1 s_1^2 + n_2 s_2^2}{n_1 + n_2} = \dfrac{35 \times 140 + 40 \times 320}{35 + 40} = 236$

$$= \frac{120-150}{\sqrt{236 \times \left(\frac{1}{35}+\frac{1}{4}\right)}} = \frac{-30}{3.556} = -8.437$$

$|Z| = 8.437$; For two-tailed test, $Z_{0.05} = 1.96$ and $Z_{0.01} = 2.58$.

As, $|Z| > Z_{0.01}$, H_0 gets rejected thus the two sample means differ significantly and hence it is concluded that the two population means differ significantly.

Chapter 12

Chi-square Test (χ^2 test)

Chi-square test is a non-parametric test as the null hypothesis does not contain any assumption regarding the population parameter. Chi-square is also written as χ^2. In this chapter two uses of chi-square test will be dealt with. They are the following-

1. Goodness of fit- To test how well an experiment fits to the theory.
2. Independence of attributes- To test whether two different attributes are independent or not.

Goodness of fit: It is a very powerful test for testing the significance of the discrepancy between theory and experiment. The test is given by Prof. Karl Pearson in 1900 and is known as "chi-square test of goodness of fit". It enables us to find if the deviation of the experiment from theory is just by chance or it is really due to the inadequacy of the theory to fit the observed data.

Let O_i, $(i = 1, 2,\ldots,n)$ is a set of observed or experimental frequencies & E_i $(i = 1,2,\ldots,n)$ is the corresponding set of expected or theoretical frequencies.

The steps involved in the test are the following:

(a) H_0:There is no significant difference between observed and expected frequencies.

(b) H_1:There is significant difference between observed and expected frequencies.

(c) Level of significance, $\alpha = 0.05$ or 0.01, usually

(d) Test statistic: Here Karl Pearsoin's Chi-square (χ^2) is used as test statistic.

Karl Pearsoin's Chi-square is given by $\sum_{i=1}^{n} \frac{(O_i - E_i)^2}{E}$

(e) From the χ^2-table, the critical value of χ^2 is obtained for $n - 1$ d.f and at α level of significance and . It is denoted as χ^2_{α}.

If cal $\chi^2 \geq \chi^2_{\alpha}$, then χ^2 is considered significant. Hence we reject H_0 and conclude that the observed frequencies differ significantly from theoretical frequencies.

If cal $\chi^2 < \chi^2$, then χ^2 is considered to be non-significant. Hence we accept H_0 and conclude that the observed frequencies do not differ significantly from theoretical frequencies.

Problem: A die is thrown for 90 times. The result of throwing the die is given below. Test at 5% and 1% level of significance whether the die is unbiased or not.

Outcome	1	2	3	4	5	6
Observed frequency	16	20	12	18	14	10

Solution:

Null hypothesis, H_0: the die is unbiased

Alternate hypothesis, H_1: The die is a biased one.

Level of significance : 5% and 1 %

Calculation of test statistic: The expected frequencies are calculated by assuming that the null hypothesis is true i.e., under H_0 the probability for each outcome will be 1/6.

So the expected frequencies for each outcome will, be $N \times p_i = 90 \times 1/6 = 15$

Outcome	1	2	3	4	5	6	
Observed frequency (O_i)	16	20	12	18	14	10	$\sum_{i=1}^{6} O_i = 90$
Expected frequencies (E_i)	15	15	15	15	15	15	$\sum_{i=1}^{6} E_i = 90$
$(O_i - E_i)^2$	1	25	9	9	1	25	
$(O_i - E_i)^2/E_i$	0.067	1.667	0.6	0.6	0.067	1.667	$\sum \frac{(O_i - E_i)^2}{E_i}$

$$\chi^2 = \sum_{i=1}^{6} \frac{(O_i - E_i)^2}{E_i} = 4.668$$

Tabulated value of χ^2 for 6 – 1 = 5 d.f at 5% level of significance, $\chi^2_{0.05} = 11.07$

Tabulated value of χ^2 for 6 – 1 = 5 d.f at 1% level of significance, $\chi^2_{0.01} = 15.08$

As Cal $\chi^2 <$ tab $\chi^2_{0.05}$ so H_0 is accepted and it is concluded that the observed frequencies do not differ significantly from the theoretical frequencies. Thus the die can be considered to be unbiased.

Problem

In an experiment on pea plants, out of 800 selected plants, 500 plants are tall and have round seeds (TTRR), 100 plants are tall and have wrinkled seeds (TTrr),

120 plants are dwarf and have round seeds (ttRR) and remaining 80 plants are dwarf and wrinkled seeds (ttrr). Test at 5% and 1% level of significance whether or not the observed data fit to the Mendel's dihybrid ratio.

Solution

Here no. of groups, $n = 4$; Number of observations, $N = 800$.

Null Hypothesis, H_0: The data fit to Mendel's dihybrid ratio, 9:3:3:1.

Alternative hypothesis, H_1: The data do not fit to Mendel's dihybrid ratio, 9:3:3:1.

Level of significance, $\alpha = 5\%$ (0.05) and 1% (0.01)

Test statistic: Since Mendel's dihybrid ratio is 9:3:3:1

Thus under H_0 (i.e., assuming H_0 to be true), P(the plant is TTRR) $= P_1 = 9/16$; P(the plant is TTrr) $= P_2 = 3/16$; P (the plant is ttRR) $= P_3 = 3/16$; P(the plant is ttrr) $= P_4 = 1/16$;

Observation	**Observed frequency (O_i)**	**Expected frequency (E_i) $E_i = N \times P_i$**	**$(O_i - E_i)^2$**	$\frac{(O_{i-E_i})^2}{E_i}$
TTRR	500	$800 \times \frac{9}{16} = 450$	2500	5.556
TTrr	100	$800 \times \frac{}{16} = 150$	2500	16.667
ttRR	120	$800 \times \frac{3}{16} = 150$	900	6.000
ttrr	80	$800 \times \frac{1}{16} = 50$	900	18.000
				$\chi^2 = \sum_{i=1}^{n} \frac{(O_i - E_i)^2}{E_i}$ $= 46.222$

Thus cal $\chi^2 = 46.222$.

Conclusion: Tabulated χ^2 at 0.05 level of significance and $4 - 1(=3)$ d.f. $= 7.81$
Tabulated χ^2 at 0.01 level of significance and $4 - 1(=3)$ d.f. $= 11.34$

Thus cal $\chi^2 >$ Tabulated χ^2 at 0.01 level of significance. So H_0 is rejected at 1% level of significance. χ^2 is considered to be significant at 1% level of significance. Thus we conclude that the data do not fit to the Mendel's dihybrid ratio.

Test of independence of attributes: Let us consider two attributes A and B, A divided into r classes $A_1, A_2, \ldots, A_r$ and B is divided into s classes $B_1, B_2, \ldots, B_s$.

Such a classification in which attributes are divided into more than two classes is known as manifold classification.

$r \times s$ contingency table: When the individuals are classified into two categories, one having 'r' classes and other having 's' classes then the table showing frequencies distributed over $r \times s$ classes is called $r \times s$ contingency table.

B \ A	A_1	A_2...	.. A_i...	. . . A_r	Row Totals
B_1	A_1B_1	A_2B_1. . .	. . A_iB_1. . . .	. . . A_rB_1	R_1
B_2	A_1B_2	A_2B_2. . .	. . A_iB_2. . . .	. . . A_rB_2	R_2
.					
.	.	.	.	.	.
.	.	.	.	.	.
.	.	.	.	.	R_j
B_j	A_1B_j	A_2B_j. . .	. . A_iB_j. . . .	. . . A_rB_j	
.					
.	.	.	.	.	.
.	.	.	.	.	.
B_s	A_1B_s	A_2B_s. . .	. . A_iB_s. . . .	. . . A_rB_s	R_s
Column totals	C_1	C_2. . .	. . C_i. . . .	. . . C_r	N

$R_1, R_2, \ldots, R_s$ are the marginal totals of the 1st row, 2nd row,..., sth row
$C_1, C_2, \ldots, C_r$ are the marginal totals of the 1st column, 2nd column,...., rth column
N = Grand total

The expected cell frequencies are calculated as: $(A_iB_j)_E = \dfrac{R_j \times C_i}{N}$

$P(A_iB_j) = P(A_i) \,.\, P(B_j)$ (Assuming H_0 to be true i.e., the attributes are independent)

$$= \frac{C_j}{N} \times \frac{R_i}{N}$$

Expected frequency, $(A_iB_j)_E = N \times P(A_iB_j) \; = N \times \dfrac{C_j}{N} \times \dfrac{R_i}{N} = \dfrac{C_j \times R_i}{N}$

Null hypothesis, H_0 : The attributes are independent.

Alternate hypothesis: H_1: The attributes are dependent

Level of significance is fixed at α which is usually 5% or 1%.

Test statistic, $\chi^2 = \sum_{i=1}^{s} \sum_{j=1}^{r} \dfrac{\left\{(A_iB_j) - (A_iB_j)_E\right\}^2}{(A_iB_j)_E}$

the degrees of freedom for $r \times s$ contingency table is $(r-1)(s-1)$

The tabulated value of χ^2 for $(r-1)(s-1)$ d.f. at α level of significance is obtained from the χ^2 table. It is denoted as χ^2_α

If cal $\chi^2 \geq \chi^2_\alpha$, then χ^2 is considered significant. Hence we reject H_0 and conclude that the the attributes are not independent.

If cal $\chi^2 < \chi^2_\alpha$, then χ^2 is considered to be non-significant. Hence we accept H_0 and conclude that the attributes are independent.

2 × 2 contingency table: When the individuals are classified into two categories w.r.t each of the two attributes then the table showing frequencies distributed over 2 × 2 classes is called 2 × 2 contingency table.

Suppose the individuals are classified according to two attributes each having two attributes. The distribution of frequencies over cells is shown in the following table:

A \ B	A_1	A_2	Row Totals
B_1	a	b	$R_1 = a + b$
B_2	c	d	$R_2 = c + d$
Column totals	$C_1 = a + c$	$C_2 = b + d$	$N = a + b + c + d$

Where R_1 and R_2 are the marginal totals of 1st row and 2nd row C_1 and C_2 are the marginal totals of 1st column and 2nd column

N = grand total

The null hypothesis H_0: the two attributes are independent.

Alternative hypothesis H_1: the two attributes are not independent.

Level of significance is fixed at α which is usually 5% or 1%.

Calculation of test statistic:

Based on H_0, the expected frequencies are calculated as follows.

$E(a) = R_1 . C_1/N$; $E(b) = R_1 . C_2/N$; $E(c) = R_2 . C_1/N$; $E(d) = R_2 . C_2/N$

Where $N = a + b + c + d$

$$\chi^2 = \frac{\{a - E(a)\}^2}{E(a)} + \frac{\{b - E(b)\}^2}{E(b)} + \frac{\{c - E(c)\}^2}{E(c)} + \frac{\{d - E(d)\}^2}{E(d)}$$

For 2 × 2 contingency table the value of χ^2 can also be obtained directly by the following formula:

$$\chi^2 = \frac{N(ad-bc)^2}{R_1 \times R_2 \times C_1 \times C_2}$$

The degrees of freedom for 2 × 2 contingency table is (2 – 1)(2 – 1) = 1

The tabulated value of χ^2 for 1 d.f. at α level of significance is obtained from the

χ^2 table. It is denoted as χ^2_α.

If cal $\chi^2 \geq \chi^2_\alpha$, then χ^2 is considered significant. Hence we reject H_0 and conclude that the attributes are not independent.

If cal $\chi^2 < \chi^2_\alpha$, then χ^2 is considered to be non-significant. Hence we accept H_0 and conclude that the attributes are independent.

Conditions for validity of chi-square test

1. The sample observations should be independent.
2. $\sum_{i=1}^{b} O_i = \sum_{i=1}^{n} E_i = N$
3. $N > 50$
4. No observed cell frequencies should be less than 5. If any of the theoretical cell frequency is less than 5, then for the application of χ^2 test, it is pooled with the preceding or succeeding frequency so that the pooled frequency is more than 5 and finally adjusted for the d.f. lost in pooling.

Yate's correction for continuity: In a 2 × 2 contingency table, the no. of d.f. is (2 – 1) × (2 – 1) = 1. If any one of the theoretical cell frequencies is less than 5, then use of pooling method for χ^2 test results in χ^2 with 0 d.f.(since 1 d.f. is lost in pooling). In such case, we apply Yate's correction for continuity. This consists in adding 0.5 to the cell frequency which is less than 5 and then adjusting for the remaining cell frequencies accordingly. The χ^2 test of goodness of fit is then applied without pooling method.

According to Yate's correction, we subtract (or add) 0.5 from *a* and *d* and add (or subtract) 0.5 to *b* and *c* so that the marginal totals are not disturbed at all. Thus corrected values of χ^2 is given as :

$$\chi^2 = \frac{N\left[\left|(a \mp 0.5)\right|(d \mp 0.5) - (b \pm 0.5)(c \pm 0.5)\right]}{R_1 \times R_2 \times C_1 \times C_2}$$

or

$$\chi^2 = \frac{\left(|ad-bc| - \frac{N}{2}\right)^2}{R_1 \times R_2 \times C_1 \div C_2}$$

Problem

The table given below gives the frequency distribution of educational qualification (Under Graduate and Graduate or above) and type of occupation (farming and non-farming) for 500 persons of a village. Test whether or not the type of occupation is independent of educational qualification.

	Farming	**Non- farming**
Under graduate	225	80
Graduate or above	125	70

Solution:

	Farming	Non-farming	**Marginal total**
Under graduate	225	80	**305**
Graduate or above	125	70	**195**
Marginal total	**350**	**150**	**Grand total, G = 500**

Null hypothesis, H_0: The two attributes i.e., educational qualification and type of occupation are independent of each other.

Alternative hypothesis, H_1: The two attributes are not independent of each other.

Level of significance, $\alpha = 5\%$ (0.05) and 1% (0.01)

Test statistic,

$$\chi^2 = \frac{500(225 \times 70 - 80 \times 125)^2}{(225+125)(125+70)(70+80)(80+225)} = 5.294$$

Conclusion: Calculated value of $\chi^2 = 5.294$

Tabulated $\chi2$ at 0.05 level of significance and (2 – 1)(2 – 1) = 1 d.f., $\chi^2_{0.05} = 3.84$

Tabulated $\chi2$ at 0.01 level of significance and (2 – 1)(2 – 1) = 1 d.f., $\chi^2_{0.01} = 6.63$

Thus, $\chi^2_{0.05} < \text{cal}\,\chi^2 < \chi^2_{0.01}$. So, H_0 is rejected at 5% level of significance. χ^2 is considered to be significant at 5% level of significance. Thus we conclude that at 5% level of significance, the two attributes i.e., educational qualification and type of occupation are not independent of each other.

Problem

60 numbers of infected animals are chosen and out of them 48 are inoculated with a vaccine The following 2 × 2 contingency table gives the effectiveness of the vaccine towards the cure of the disease.

	Inoculated	Not inoculated
Survived	40	2
Not survived	8	10

Solution

	Inoculated	Not inoculated	**Marginal total**
Survived	40	2	**42**
Not survived	8	10	**18**
Marginal total	**48**	**12**	**Grand total, G = 60**

Null hypothesis, H_0 : The two attributes are independent of each other.

Alternative hypothesis, H_1: The two attributes are not independent of each other.

Level of significance, $\alpha = 5\%$ (0.05) and 1% (0.01). As one of the cell frequency is less than 5,

So, test statistic, $\chi^2 = \dfrac{60\left(|40\times10-2\times8|-\dfrac{60}{2}\right)^2}{(40+8)(8+10)(10+2)(2+40)} = 17.267$

Conclusion: Tabulated χ^2 at 0.05 level of significance and (2 – 1)(2 – 1) = 1 d.f., $\chi^2_{0.05} = 3.84$

Tabulated χ^2 at 0.01 level of significance and (2 – 1)(2 – 1) = 1 d.f, $\chi^2_{0.01} = 6.63$.

Calculated $\chi^2 >$ Tabulated $\chi^2_{0.01}$ So, H_0 is rejected at 1% level of significance. χ^2 is considered to be significant at 1 % level of significance. Thus we conclude that at 1% level of significance the two attributes, i.e., vaccinating and chance of survival are not independent of each other.

Chapter 13

Design of Experiments

An experiment is a systematic process or activity which leads to collection of information on certain objects to give an answer to the objectives of the researcher.

The experiment in which certain absolute values like average, correlation coefficient, median, mode, etc are worked out to describe the population is known as **absolute experiment**.

The experiment in which the effects of different objects under consideration are compared is known as **comparative experiment**.

The variable whose change we wish to study is called as **Response variable**. For example- yield, mortality rate of insects etc.

The variable whose effect on the response variable we wish to study is termed as **Factor**. For example - variety, fertilizer, insecticide etc.

Here **factor** is the independent variable and **response variable** is the dependent variable.

The various levels of a factor whose effect on the response variable we wish to study are known as **Treatment**.

Treatment is also defined as the different objects under comparison in a comparative experiment. For example, Different varieties of a crop, different doses of fertilizers, different doses of insecticides, etc.

Experimental material- The object/place where the experiment is conducted i.e., the treatments are applied is known as experimental material. For example, agricultural field, insects etc.

Experimental unit- The smallest unit of the experimental material where the treatment is applied is known as experimental unit.

For example, plot of a field, an individual insect, etc.

Let us consider some examples to make the meaning of the terms more clear

Suppose we want to study whether the yield of different varieties of a crop is significantly different or not. Here **variety** is the factor. **Yield** is the response variable.

Different varieties are the **treatments**. The **field** in which the varieties are grown is the **experimental material.** The various plots of the field are known as the **experimental units.**

Suppose we want to study whether different doses of an insecticide have significantly different effect on the mortality rate of insects. Here **insecticide** is the **factor. Mortality rate** of insects is the **response variable.** Different doses of insecticide are the **treatments**. Insects are the **experimental material**. An individual insect is the **experimental unit**.

The responses from the experimental units differ from each other, this difference is called error. This error is also known as variation. It is attributed to two causes:-

(i) Assignable cause of Variation

This variation/error arises because the different units are allocated different treatments. It is called between treatment variations.

(ii) Chance cause of Variation

This is the variation in the experimental units receiving the same treatment. These variations may arise due to the inherent differences between the experimental units or due to inaccuracies in the physical conduct of the experiment. It is called within treatment variation.

Extraneous factors

The factors which are responsible for variation in the responses of the experimental units receiving the same treatment are known as extraneous factors. Examples of such factors are soil heterogeneity, climatic factors, genetic differences, moisture content of the plots etc.

Experimental error

The difference in response of experimental units receiving the same treatment is known as experimental error. Experimental error is due to extraneous factor. It is beyond human control. We can only take steps to minimize the experimental error by adopting certain techniques.

Design of experiment

The choice of treatments, the method of assigning treatments to experimental units and arrangement of experimental units in various patterns to suit the requirement of particular problem, are combinedly known as design of experiment.

Basic principles of design of experiment

The techniques that we adopt to reduce the experimental error so as to increase the precision/accuracy of the experiment are known as principles of design of experiment. There are three principles:

Replication: The repeated application of the treatments under investigations is known as **replication.** Two major functions of replication-

(a) to provide an estimate of experimental error.

(b) to reduce the experimental error.

The number of replications required in an experiment depends on the factors like cost of the test material, variability of the test material, desired level of accuracy, the expected size of mean differences in the response, and the error variance.

Formula for getting the required no. of replications is given by, $r = \frac{2 \times s^2 t_\alpha^2}{d^2}$

where t = table value of 't' for a given level of significance and degrees of freedom.

d = The difference between the means of two treatments.

s^2 = Error variance (taken from similar experiments conducted earlier).

r = Number of replications.

In practice, it may not be possible to know s^2. So, an approximate rule to fix the number of replications is that the no. of replications should be such that the error degrees of freedom should not be less than 12.

Randomization

It is the technique in which all the treatments have equal chances of being allocated to different units. The actual procedure of randomization will vary according to the design we adopt. The major function of randomization is to provide a valid estimate of experimental error.

Local control

Grouping of homogenous experimental units into blocks in order to reduce the experimental error is called local control. The major function of local control is to reduce the experimental error. Local control can be accomplished by,

- Uniformity trial
- Examination of the results of actual experiments

Thus it can be concluded that,

To have a valid estimate of experimental error, the principles of replication and randomization are used.

To reduce the experimental error, the principles of replication and local control are used.

Analysis of variance (ANOVA)

We have read earlier in chapter 10 that F-test is also used to test the significance of difference between more than two independent sample means. This test is also known as analysis of variance (ANOVA) test. To test the significance of more than two means, we use ANOVA test. The technique was developed by R.A. Fisher.

Definition: The analysis of variance is the systematic algebraic procedure of decomposing the overall variation in the responses observed in an experiment into different components.

Total variation in the response variable = Variation between treatment variation + Variation within treatment

Thus, Total variance = Between treatment variance + Within treatment variance

Thus the overall variance is divided into at least two components. The between treatment variation can also be subdivided into two or more components. The within treatment variation can also be subdivided into two or more components in case of more than one factor ANOVA.

ANOVA dealing with one factor is known as one-way ANOVA.

ANOVA dealing with two factors is known as two-way ANOVA.

ANOVA dealing with three factors is known as three-way ANOVA.

ANOVA dealing with multiple factors is known as Multiple ANOVA or MANOVA

One-way ANOVA: Suppose we have a factor 'A' having '*t*' levels, each with equal replications '*r*'. Let their responses be denoted by Y_{ij}; where Y_{ij} is the response of jth replication of ith level of the factor A.

The result may be tabulated as:

Various level of factor A →		1	2	...*i*	...*t*
Replication ↓	1	Y_{11}	Y_{21} ...	...Y_{i1}...	...Y_{t1}
	2	Y_{12}	Y_{22} ...	...Y_{i2}...	...Y_{t2}
	.	.	.	.	.
	.	.	.	.	.
	.	.	.	.	.
	j	Y_{1j}	Y_{2j} ...	...Y_{ij}...	...Y_{tj}
	.	.	.	.	.
	.	.	.	.	.
	r	Y_{1r}	Y_{2r} ...	...Y_{ir}...	...y_{tr}
Total of ith level of factor *A*		A_1	A_2 ...	....A_i	A_t

One-way ANOVA model: $y_{ij} = \mu + a_i + e_{ij,}$ Where, y_{ij} = response of jth replication of ith level of factor A; a_i = effect of ith level of factor A; e_{ij} = error effect; μ = general mean effect.

Meaning of general mean effect: If there would have been no difference in the levels of factor A and no experimental error, then the value of the response variable for each experimental unit would have been μ.

Meaning of effect of ith level of factor A: If there would have been no experimental error, then the value of the response variable for each experimental unit receiving ith level of factor A would have been a_i.

Effect of ith level of factor A, $a_i = \overline{A}_i - \mu$; where, $\overline{A}_i$ is the mean of ith level of factor A.

Error effect is the effect due to extraneous factors.

Grand total, $G = \sum_{i=1}^{t} \sum_{j=1}^{r} Y_{ij}$; Correction factor, $GF = \dfrac{G^2}{t \times r}$.

Total of ith level of factor A, $A_i = \sum_{j=1}^{r} Y_{ij}$; Mean of ith level of factor A, $\overline{A}_i = \dfrac{\sum_{j=1}^{r} Y_{ij}}{r}$;

Calculation of Sum of Squares

Total sum of squares, $S_T^2 = \sum_{i=1}^{t} \sum_{j=1}^{r} Y_{ij}^2 - CF$; $\sum_{i=1}^{t} \sum_{j=1}^{r} y_{ij}^2$ is called raw sum of squares

$$\text{Sum of squares due to } i^{th} \text{ level of factor } A,\ S_A^2 = \frac{\sum_{i=1}^{t} A_i^2}{r} - \text{CF}$$

Error sum of squares, $S_E^2 = S_T^2 - S_A^2$

Calculation of Mean Squares

Mean square due to i^{th} level of factor A (AMS), $s_A^2 = \dfrac{S_A^2}{t-1}$

Error mean square (EMS), $s_E^2 = \dfrac{s_E^2}{(t)(r-1)}$

Null hypothesis: $H_0(A)$: $a_1 = a_2 = \ldots = a_t$ (i.e., effect of all levels of factor A are identical).

Alternative hypothesis: $H_1(A)$: $a_1 \neq a_2 \neq \ldots \neq a_t$ (i.e., effect of at least any two levels of factor A differ significantly).

The null and alternative hypothesis can also be framed in the following manner:

Null hypothesis: $H_0(A)$: $\overline{A_1} = \overline{A_2} = \cdots = \overline{A_t}$ (i.e., mean of all levels of factor A are identical)

Alternative hypothesis: $H_1(A)$: $\overline{A_1} \neq \overline{A_2} \neq \cdots \neq \overline{A_t}$ (i.e., mean of at least any two levels of factor A differ significantly).

This is because, effect of ith level of factor A, $a_i = \overline{A_i} - \mu$. So, $a_1 - a_2 = \overline{A_1} - \overline{A_2}$

Level of significance = ±; α =0.05 or 0.01 (usually);

Test statistic, $F_A = \dfrac{\text{AMS}}{\text{EMS}}$

ANOVA table for one-way ANOVA

Sources of variation	d.f.	Sum of squares	Mean squares	F
i^{th} level of factor A	$t - 1$	S_A^2	S_A^2	$F_A = S_A^2/S_E^2$
Error	$n - t = rt - t = t(r - 1)$	S_E^2	S_E^2	
Total	$n - 1$	S_T^2		

Tabulated value of F is obtained from F table for $(t - 1)$, $(r - 1)(t)$ d.f. at α level of significance.

If Calculated $F_A \geq$ tabulated $F\alpha$ for $(t - 1)$, $(r - 1)(t)$ d.f., then F_A is considered significant. We thus reject H_0 at α L.S. and conclude that the effect (mean) of at least any two levels of factor A differ significantly.

If Calculated $F_A <$ tabulated $F\alpha$ for $(t - 1)$, $(r - 1)(t)$ d.f then F is considered insignificant. We thus accept H_0 at α L. S. and conclude that the effect (mean) of different levels of factor A is identical.

Two way ANOVA: Here two factors are involved. Suppose we have 't' levels of factor 'A' and 's' levels of factors 'B' every combination of the two factors are involved once. This is called two-way ANOVA with one observation per cell.

Let 'y_{ij}' he the response of 'i^{th}' level of factor A and 'j^{th}' level of factor B.

Two-way ANOVA model: $y_{ij} = \mu + a_i + b_j + e_{ij}$

Where, y_{ij} = response of jth level of factor B and i^{th} level of factor A; a_i = effect of ith level of factor A; b_j = effect of jth level of factor B; e_{ij} = error effect; μ = general mean effect.

Meaning of general mean effect: If there would have no difference in the levels of factors A and B and no experimental error, then the value of the response variable for each experimental unit would have been μ.

Meaning of effect of ith level of factor A: If there would have been no difference in levels of factor B and no experimental error, then the value of the response variable for each experimental unit receiving i^{th} level of factor A would have been a_i.

Effect of ith level of factor A, $a_i = \overline{A_i} - \mu$; where $\overline{A_i}$ is the mean of ith level of factor A.

Meaning of effect of jth level of factor B: If there would have been no difference in levels of factor A and no experimental error, then the value of the response variable for each experimental unit receiving j^{th} level of factor B would have been b_j.

Effect of jth level of factor B, $b_j = \overline{B_j} - \mu$; where $\overline{B_j}$ is the mean of jth level of factor B.

Error effect is the effect due to extraneous factors.

The result is tabulated as:

Various level of factor A →		1	2	...i	...t	Total of jth level of factor B
Various level of factor B ↓	1	Y_{11}	$Y_{21}\cdots$	$\ldots Y_{i1}\cdots$	$\ldots Y_{t1}$	B_1
	2	Y_{12}	$Y_{22}\cdots$	$\ldots Y_{i2}\cdots$	$\ldots Y_{t2}$	B_2
	.	.	.	.	.	.
	.	.	.	.	.	.
	.	.	.	.	.	.
	j	Y_{1j}	$Y_{1j}\cdots$	$\ldots Y_{ij}\cdots$	$\ldots Y_{tj}$	B_j
	.	.	.	.	.	.
	.	.	.	.	.	.
	.	.	.	.	.	.
	s	Y_{1s}	$Y_{2s}\cdots$	$\ldots Y_{is}\cdots$	$\ldots y_{ts}$	B_s
Total of ith level of factor A		A_1	$A_2\cdots$	$\ldots A_i$	A_t	G

Grand total, $G = \sum_{i=1}^{t}\sum_{j=1}^{s} Y_{ij}$ Correction factor, $CF = \frac{G^2}{t \times s}$

Total of ith level of factor A, $A_i = \sum_{j=1}^{s} Y_{ij}$; ; Total of jth level of factor B, $B_j = \sum_{i=1}^{t} Y_{ij}$;

Calculation of Sum of Squares

Total sum of squares, $S_T^2 = \sum_{i=1}^{t}\sum_{j=1}^{s} Y_{ij}^2 - CF$; $\sum_{i=1}^{t}\sum_{j=1}^{s} Y_{ij}^2$ is called raw sum of squares.

Sum of squares due to i^{th} level of factor A, $S_A^2 = \frac{\sum_{i=1}^{t} A_i^2}{s} - CF$

Sum of squares due to jth level of factor B, $S_B^2 = \frac{\sum_{j=1}^{s} B_j^2}{t} - CF$

Error sum of squares, $S_E^2 = S_T^2 - S_A^2 - S_B^2$

Calculation of Mean Squares

Mean square due to $_i^{th}$ level of factor $A\,(AMS)$, $s_A^2 = \frac{S_A^2}{t-1}$

Mean square due to j^{th} level of factor $B\,(BMS)$, $s_B^2 = \frac{S_B^2}{s-1}$

Error mean square (EMS), $s_E^2 = \frac{S_E^2}{(t-1)(s-1)}$

For factor A

Null hypothesis: $H_0(A)$: $a_1 = a_2 = \ldots = a_t$ (i.e., effect of all levels of factor A are identical)

Alternative hypothesis: $H_1(A)$: $a_1 \neq a_2 \neq \ldots \neq a_t$(i.e., effect of at least any two levels of factor A differ significantly).

The null and alternative hypothesis can also be framed in the following manner:

Null hypothesis: $H_0(A)$: $\overline{A_1} = \overline{A_2} =, \cdots = \overline{A_t}$ (i.e., mean of all levels of factor A are identical)

Alternative hypothesis: $H_1(A)$: $\overline{A_1} \neq \overline{A_2} \neq, \cdots \neq \overline{A_t}$ (i.e., mean of at least any two levels of factor A differ significantly).

This is because, effect of ith level of factor A, $a_i = \overline{A_i} - \mu$; So, $a_1 - a_2 = \overline{A_1} - \overline{A_2}$

For factor B

Null hypothesis: $H_0(B)$: $b_1 = b_2 = \ldots = b_t$ (i.e., effect of all levels of factor B are identical).

Alternative hypothesis: $H_1(B)$: $b_1 \neq b_2 \neq \ldots \neq b_t$(i.e., effect of at least any two levels of factor B differ significantly).

The null and alternative hypothesis can also be framed in the following manner:

Null hypothesis: $H_0(B)$: $\overline{B_1} = \overline{B_2} = \cdots = \overline{B_S}$ (i.e., mean of all levels of factor B are identical)

Alternative hypothesis: $H_1(B)$: $\overline{B_1} \neq \overline{B_2} \neq \cdots \neq \overline{B_S}$ (i.e., mean of at least any two levels of factor **B** differ significantly).

This is because, effect of jth level of factor **B**, $b_j = \overline{B_j} - \mu$;

So, $= b_1 - b_2 = \overline{B_1} - \overline{B_2}$

Level of significance is which is usually, $\alpha = 0.05$ or 0.01 (usually)

Anova table for Two-way ANOVA

Sources of variation	d.f.	Sum of squares	Mean squares	F
i^{th} level of factor A	$t - 1$	S_A^2	s_A^2	$F_A = s_A^2/s_E^2$
j^{th} level of factor B	$s - 1$	S_B^2	s_B^2	$F_B = s_B^2/s_B^2$
Error	$(t - 1)(s - 1)$	S_E^2	s_E^2	
Total	$(t \times s) - 1$	S_T^2		

Test statistic, $F_A = \dfrac{\text{AMS}}{\text{EMS}}$ or $\dfrac{s_A^2}{s_E^2}$; $F_B = \dfrac{\text{BMS}}{\text{EMS}}$

For factor *A*, tabulated *F* value is obtained from *F* table for $(t - 1)$, $(t - 1)(s - 1)$ d.f. at α level of significance.

For factor *B*, tabulated *F* value is obtained from *F* table for $(s - 1)$, $(t - 1)(s - 1)$ d.f. at α level of significance.

If Calculated $F_A \geq$ tab F_α for $(t - 1)$, $(t - 1)(s - 1)$ d.f., then F_A is considered significant. We thus reject $H_{0(A)}$ at α level of significance and conclude that the means (i.e., effect) of at least two levels of factor *A* differ significantly, otherwise the mean (i.e., effect) of all levels of factor A are identical.

If calculated $F_B \geq$ tab F_α for $(s - 1)$, $(t - 1)(s - 1)$ d.f., then F_B is considered significant. We thus reject $H_{0(B)}$ at α level of significance and conclude that the means (i.e., effect) for at least two levels of factor *B* differ significantly, otherwise the mean (i.e., effect) of all levels of factor B are identical.

When the treatment consists of different level of a single factor, it is known as factor experiment. CRD, RCBD and LSD are all single factor experimental design.

Types of ANOVA model

1. Fixed effect model or model I: The treatments which are to be replicated are fixed.
2. Random effect model or model II or variance component model: The treatment which are to be replicated, are selected randomly.
3. Mixed effect model or model III: Some treatments are fixed and some treatments are selected randomly.

Single factor design: There are three different single factor desings:

1. CBD (Completely Randomized Deisgn)
2. RCBD (Randomised Complete Block Design)
3. LSD (Latin Square Design)

Completely Randomized Design (CRD)

It is the basic single factor experimental design. RCBD and LSD stem from it by imposing restrictions upon the allocation of the treatments within the experimental material.

Definition of CRD

When the treatments are arranged randomly over the whole of a previously determined set of experimental units, the design is known as CRD. No. of replications in CRD may or may not be equal for all the treatments.

Layout of CRD

The placement of the treatments on the experimental units along with the arrangement of experimental units is known as the layout of an experiment. Let '*t*' number of treatments (named as T_1, T_2, $T_3 \ldots T_t$) be replicated '*r*' times each. Thus number of plots/experimental units required, $n = r \times t$.

Then entire experimental material is divided into n ($= r \times t$) number of units which are numbered from 1 to n in a serpentine manner. Then '*n*' distinct random numbers are selected from random number tables and they are ranked in ascending order. These ranks correspond to the plot or experimental unit number of the experimental material. The first set of '*r*' units are allocated to treatment T_1, the second set of '*r*' units are allocated to the treatment T_2 and so on.

Statistical analysis of CRD

The ANOVA model for CRD is $y_{ij} = \mu + t_i + e_{ij}$

Where, y_{ij} = response of jth replication of ith treatment; t_i = i^{th} treatment effect; e_{ij} = error effect.

μ = general mean effect.

Meaning of general mean effect: If there would have no difference in the treatments and no experimental error, then the value of the response variable for each experimental unit would have been μ.

Meaning of ith treatment effect: If there would have no no experimental error, then the value of the response variable for each experimental unit of ith treatment would have been t_i.

i^{th} treatment effect, $t_i = \overline{T_i} - \mu$ where $\overline{T_i}$ is the ith treatment mean

Error effect is the effect due to extraneous factors.

The tabular arrangement of data (for equal replications)

	1	2	i....	T
1	Y_{11}	Y_{21}...	Y_{i1} ...	Y_{t1}
2 • •	Y_{12} • •	Y_{22}... • •	Y_{i2} ... • •	Y_{t2} • •
j • •	Y_{1j} • •	Y_{2j} • •	Y_{ij} ... • •	Y_{tj} • •
r	Y_{1r}	Y_{2r}	Y_{ir} ...	Y_{tr}
i^{th} treatment total	T_1	T_2.....	T_i ...	T_t

The tabular arrangement of data (for unequal replications)

	1	2	i....	t
1	Y_{11}	Y_{21}...	Y_{i1} ...	Y_{t1}
2 • •	Y_{12} . 	Y_{22}... • •	Y_{i2} ... • •	Y_{t2} • •
J • •	Y_{1j} • •	Y_{2j} • •	Y_{ij} ... • •	Y_{tj} • •
r_i	Y_{1r1}	Y_{2r2}	Y_{iri} ...	Y_{trt} • •
i^{th} treatment total	T_1	T_2.....	T_i ...	T_t

Grand total, $G = \sum_{i=1}^{t} \sum_{j=1}^{r} Y_{ij}$ (for equal replications);

Grand total, $G = \sum_{i=1}^{t} \sum_{j=1}^{r_i} Y_{ij}$ (for unequal replications);

Correction Factor, $CF = \dfrac{G^2}{r \times t}$

Calculation of Sum of Squares

1. Total sum of squares (TSS), $S_T^2 = \sum_{i=1}^{t} \sum_{j=1}^{r} Y_{ij}^2 - CF$ (for equal replications);

   ($\sum_{i=1}^{t} \sum_{j=1}^{r} Y_{ij}^2$ is called raw sum of squares)

   Total sum of squares (TSS), $S_T^2 = \sum_{i=1}^{t} \sum_{j=1}^{r_i} Y_{ij}^2 - CF$ (for unequal replications);

   ($\sum_{i=1}^{t} \sum_{j=1}^{r_i} Y_{ij}^2$ is called raw sum of squares;)

   Correction Factor, $CF = \dfrac{G^2}{n}$

   '*n*' is the total no. of observations; $n = r \times t$ (for equal replications);

   $n = \sum_{i=1}^{t} r_i$ (for unequal replications)

2. Treatment sum of squares (T_r.S.S), $S_t^2 = \frac{\sum_{i=1}^{t} T_i^2}{r} - CF$; where, T_i is the ith treatment total i^{th} treatment total, $T_i = \sum_{j=1}^{r} Y_{ij}$ (for equal replications); i^{th} treatment total, $T_i = \sum_{j=1}^{r_i} Y_{ij}$ (for unequal replications);

3. Error sum of squares (SSE or ESS), $S_E^2 = S_T^2 - S_t^2$

Calculation of mean squares → Mean square $\frac{\text{Sum of squares}}{\text{corresponding d.f.}}$

1. Treatment mean squares(T.M.S), $s_t^2 = \frac{S_t^2}{t-1}$

2. Error mean square(E.M.S), $s_E^2 = \frac{S_E^2}{(n-t)}$, Here, Number of observations, $n = \sum_{i=1}^{t} r_i$ (for unequal replications), $n = r \times t$ (for equal replications)

So, for equal replications, Error mean square(E.M.S), $s_E^2 = \frac{S_E^2}{(r-1) \times t}$;

Null hypothesis, $H_0: t_1 = t_2 = \ldots = t_t$ (i.e., the treatment effects are identical)

Alternative hypothesis, $H_1: t_1 \neq t_2 \neq \ldots \neq t_t$ (i.e., at least any two treatment effects differ significantly)

The null and alternative hypothesis can also be framed in the following manner:

Null hypothesis, $H_0: \overline{T_1} = \overline{T_2} = \cdots = \overline{T_t}$ (i.e., the treatment means are identical)

Alternative hypothesis, $H_1: \overline{T_1} \neq \overline{T_2} \neq \cdots \neq \overline{T_t}$ (i.e., at least any two treatment means differ significantly).

This is because, i^{th} treatment effect, $t_i = \overline{T_i} - \mu$; So, $t_1 - t_2 = \overline{T_1} - \overline{T_2}$

Level of significance, $\alpha = 0.05(5\%)$ or $0.01(1\%)$

ANOVA table for CRD

Sources of variation	d.f.	Sum of squares	Mean squares	F
Treatment	$t-1$	S_t^2	s_t^2	$F = s_t^2/s_E^2$
Error	$n-t$	S_E^2	s_E^2	
Total	$n-1$	S_T^2		

Test statistic, $F_T = s_t^2/s_E^2 = \text{TMS/EMS}$

Tabulated value of F_T is obtained from F-table at $(t-1, n-t)$ d.f. for α level of significance.

If calculated $F_T \geq$ tab F_α at $(t-1, n-t)$ d.f. then F_T is considered significant. We thus reject H_0 and conclude that the treatment effect (i.e., the treatment means) differ significantly.

If calculated $F_T <$ tab F_α at $(t-1, n-t)$ d.f., then F_T is considered insignificant. We thus accept H_0 and conclude that the treatment effect (i.e., the treatment means) do not differ significantly.

If F_T is significant, we go for C.D. test to find out actually which pair of treatment means differ significantly.

Critical difference (C.D.) test or Least Significant Difference (L.S.D.) test

Critical difference test, the C.D. test is a form of t-test. Its formula is given by,

C.D.= t_α at error d.f × SE(d); where, $d \rightarrow$ difference between the treatment means

$$SE(d) = \sqrt{EMS \times \left(\frac{1}{r_i} + \frac{1}{r_k}\right)}$$

r_i is the no. of replications for i^{th} treatment; r_k is the no. of replications for k^{th} treatment; In case of equal replications

$$SE(d) = \sqrt{\frac{2 \times EMS}{r}}$$

t-tabulated t value for a specified level of significance and error degree of freedom

Two treatment means are declared significantly different at a specified level of significance if their difference exceeds the calculated CD value; otherwise they are not significantly different i.e., said to be at par. The treatment means which are at par are connected by a bar or line.

Critical difference or least significant difference is the least value for which the difference between the means of two level of a factor (treatment) is declared to be significant

Advantage of CD test: It is easy to calculate and it provides a single value for testing all difference

Limitation of CD test:

1. It is easy to used only if F is significant.
2. For unequal replication there are different CD values for comparison tedious if there are large no. of treatment.

Merits of CRD

1. Chief advantage is that the number of replications may be varied from treatment. Because of this flexibility, all the available experimental material can be utilised without any wastage.
2. Statistical analysis is very simple.
3. Even when some observations are missing, the analysis can be carried out by deleting that observation because this design allows unequal no. of replications.
4. It provides the maximum no of d.f. for the estimation of experimental error.

Demerits of CRD

The major disadvantage of CRD is that it is less accurate than other design i.e., its efficiency is less. When no. of treatments are large, a relatively large amount of experimental material must be used. Hence, the heterogeneity of experimental will be increased. This will result in increased experimental error & reduced precision.

CRD can be used only when the experimental material is homogenous. Hence CRD is not preferred in agricultural fields are highly heterogeneous.

Uses of CRD

1. Used in green house studies
2. Used in laboratory experiments

In CRD, only Replication & Randomization are applied. In CRD principle of local control is not applied.

Problem

In a pot experiment laid out in CRD, the yield (in g per pot) of 5 varieties of a plant with 3, 5, 4, 5, 3 replications respectively are given below:

Varieties → Replication ↓	A	B	C	D	E
1	527	366	522	354	522
2	428	395	596	464	564
3	469	468	579	454	575
4	-	502	618	450	-
5	-	518	-	474	-

Test whether the varieties differ significantly with respect to yield or not.

Solution:

Varieties → Replication ↓	A	B	C	D	E
1	527	366	522	354	522
2	428	395	596	464	564
3	469	468	579	454	575
4	-	502	618	450	-
5	-	518	-	474	-
Treatment Total (T_i)	1424	2249	2315	2196	1661
Treatment Mean ($\overline{T_i}$)	474.6667	449.8	578.75	439.2	553.6667

Let y_{ij} denote the yield of ith treatment for j^{th} replication. $i = 1, 2, 3, 4, 5$; $r_1 = 3, r_2 = 5, r_3 = 4, r_4 = 5, r_5 = 3$

Number of treatments, $t = 5$; No. of observations, $n = \sum_{i=1}^{5} r_i = 3+5+4+5+3 = 20$

Grand total, $G = \sum_{i=1}^{t}\sum_{j=1}^{r_i} y_{ij} = 527 + 366 + \cdots + 518 + 474 = 9845$

Correction factor, $CF = \dfrac{G^2}{n} = \dfrac{(9845)^2}{20} = 4846201$

Treatment Total, $T_i = \sum_{j=1}^{r_i} y_{ij}$; Treatment Mean, $\overline{T_i} = \dfrac{T_i}{r_i}$

Calculation of sum of squares

1. Total sum of squares,

$$S_T^2 = \sum_{i=1}^{t}\sum_{j=1}^{r_i} y_{ij}^2 - CF = 4950181 - 4846201 = 103979.75$$

2. Treatment Sum of Squares, $S_{TR}^2 = \dfrac{\sum_{i=1}^{t} T_i^2}{r_i} - C.F.$

$$= \frac{1424^2 + 2249^2 + \cdots + 1661^2}{4} - 4846201 = 4911455 - 4846201 = 65254.07$$

3. Error Sum of Squares, $S_E^2 = S_T^2 - S_{TR}^2 = 103979.8 - 65265.07 = 38725.68$

Calculation of Mean Squares: Mean Square $= \dfrac{SS}{df}$; SS – Sum of Squares; df – degrees of freedom

1. Treatment Mean Square, $\text{TMS} = \dfrac{S_{TR}^2}{t-1} = \dfrac{65254.07}{5-1} = 16313.52$

2. Error Mean Square, $\text{EMS} = \frac{S_E^2}{n-t} = \frac{38725.68}{20-5} = 2581.712$

3. Null hypothesis, H_0: $\overline{T_1} = \overline{T_2} = \overline{T_3} = \overline{T_4} = \overline{T_5}$;

   Alternative hypothesis, H_1: $\overline{T_1} \neq \overline{T_2} \neq \overline{T_3} \neq \overline{T_4} \neq \overline{T_5}$

Level of significance, $\alpha = 0.05$ (5%) and 0.01 (1%)

Test Statistic F for comparing the treatment means,

$$F_T = \frac{TMS}{EMS} = \frac{16313.52}{2581.712} = 6.32$$

Conclusion: Calculated Value of F for comparing the treatment means, $F_T = 6.32$

Tabulated value of F at 4, 15 d.f. for 0.05(5%) Level of significance, $F_{0.05} = 3.06$

Tabulated value of F at 4, 15 d.f. for 0.01(1%) Level of significance, $F_{0.01} = 4.89$

As $F_T > F_{0.01}$, H_0 is rejected at 0.01 (1%) Level of Significance. F_T is considered significant at 0.01 (1%) Level of Significance Thus it is concluded that the treatment means differ significantly at 1 % level of Significance.

ANOVA Table for CRD

Sources of variation	Degrees of Freedom (d.f.)	Sum of Squares (SS)	Mean Squares (MS)	F value		
				Calculated F	Tabulated F	
					0.05 LS	0.01 LS
Treatments	5 – 1 = 4	$S_{TR}^2 = 65254.07$	TMS = 16313.52	F_T = TMS/EMS = 6.32**	3.06	4.89
Error	5(4 – 1) = 15	$S_E^2 = 38725.68$	EMS = 2581.712			
Total	5 × 4 – 1 = 19	$S_T^2 = 103979.75$				

As F_T is significant i.e., the treatment means differ significantly, C.D. (Critical Difference) test is to be done to know actually which pair of treatment means differ significantly.

C.D. $= t_1 \times SE(d)$; t_1 – tabulated value of t for 0.05 (5%) Level of Significance at error degrees of freedom.

$SE(d)$ – Standard error of difference between treatment means.

$$SE(d) = \sqrt{EMS\left(\frac{1}{r_i} + \frac{1}{r_j}\right)}$$

Where, r_i and r_j are the number of replications for i^{th} and j^{th} treatments respectively.

d is the difference between means of ith and j^{th} treatments i.e., $d = \overline{T}_i - \overline{T}_j \forall i, j = 1$ to 5 and $i \neq j$,

$$SE(\overline{T}_3 - \overline{T}_5) = \sqrt{2581.713 \times \left(\frac{1}{4} + \frac{1}{3}\right)} = \sqrt{1505.999} = 38.807;$$

$$\text{C.D.} = 2.131 \times 38.807 = 82.7$$

In the similar manner the SE(d) and the corresponding C.D. values are obtained for all the possible pairs of treatment means

The treatment means arranged in descending order as: $\overline{T}_3 > \overline{T}_5 > \overline{T}_1 > \overline{T}_2 > \overline{T}_4$
$\overline{T}_3$ (578.75); $\overline{T}_5$ (553.67); $\overline{T}_1$ (474.67); $\overline{T}_2$ (449.8.25); $\overline{T}_4$ (439.20).

Difference	$\overline{T}_3 - \overline{T}_5 =$	$\overline{T}_3 - \overline{T}_1 =$	$\overline{T}_3 - \overline{T}_2 =$	$\overline{T}_3 - \overline{T}_4 =$	$\overline{T}_5 - \overline{T}_1 =$	$\overline{T}_5 - \overline{T}_2 =$	$\overline{T}_5 - \overline{T}_4 =$	$\overline{T}_1 - \overline{T}_2 =$	$\overline{T}_1 - \overline{T4} =$	$\overline{T}_2 - \overline{T}_4 =$
	25.08	104.08	128.95	139.55	79.00	103.87	114.47	24.87	35.47	10.60
C.D. value	82.7	82.7	72.63	72.63	88.41	79.07	79.07	79.07	79.07	68.48
S/NS	*NS*	*S*	*S*	*S*	*NS*	*S*	*S*	*NS*	*NS*	*NS*

The result of CD test is: $\overline{T}_3$ $\overline{T}_5$ $\overline{T}_1$ $\overline{T}_2$ $\overline{T}_4$
(ab) (bc) (cd) (de) (df)

Conclusion: The yield of Variety *C* is the highest which is at par with variety *E* but significantly higher than the remaining three varieties *A*, *B* & *D*. Variety *E* gives second highest yield which is at par with variety *A* but significantly higher than varieties *B* & *D*. Variety D gives the lowest yield which is at par with Varieties *A* & *B*.

Problem

In a pot experiment laid out in CRD with 4 replications each the yield (in g per pot) of 5 varieties of a plant are given below:

Varieties … Replication ↓	*A*	*B*	*C*	*D*	*E*
1	557	376	512	344	532
2	458	385	586	444	554
3	469	458	569	454	555
4	461	502	608	450	566

Test whether or not the varieties differ significantly with respect to yield.

Solution

Varieties → (Treatments) Replication↓	*A*	*B*	*C*	*D*	*E*
1	557	376	512	344	532
2	458	385	586	444	554
3	469	458	569	454	555
4	461	502	608	450	566
Treatment Total (T_i)	1945	1721	2275	1692	2207
Treatment Mean ($\overline{T}_i$)	486.25	430.25	568.75	423	551.75

Let y_{ij} denote the yield of ith treatment for j^{th} replication. $i = 1, 2, 3, 4, 5; j = 1, 2, 3, 4$

Number of treatments, $t = 5$; No. of replications, $r = 4$; Number of observations, $n = r \times t = 4 \times 5 = 20$

Grand total, $G = \sum_{i=1}^{5}\sum_{j=1}^{4} y_{ij} = 557 + 376 + \cdots + 450 + 566 = 9840$;

Correction factor, $C.F. = \dfrac{G^2}{n} = \dfrac{(9840)^2}{20} = 4841280$

Treatment Total, $T_i = \sum_{j=1}^{4} y_{ij}$; Treatment Mean , $\overline{T}_i = \dfrac{T_i}{4}$;

Calculation of sum of squares

4. Total sum of squares, $S_T^2 = \sum_{i=1}^{5}\sum_{j=1}^{4} y_{ij}^2 - CF = 4945238 - 4841280 = 103958$

5. Treatment Sum of Squares, $S_{TR}^2 = \dfrac{\sum_{i=1}^{5} T_i^2}{4} - C.F.$

$$= \frac{1945^2 + 1721^2 + \cdots + 2207^2}{4} - 4841280 = 4913551 - 4841280 = 72271$$

6. Error Sum of Squares, $S_E^2 = S_T^2 - S_{TR}^2 = 103958 - 72271 = 31687$

Calculation of mean squares: mean square $= \dfrac{SS}{df}$; SS – Sum of Squares; df – degrees of freedom

7. Treatment Mean Square, $TMS = \dfrac{S_{TR}^2}{t-1} = \dfrac{72271}{5-1} = 18067.75$

8. Error Mean Square, $EMS = \dfrac{S_E^2}{t(r-1)} = \dfrac{31687}{5(4-1)} = 2112.47$

Null hypothesis, H_0**:** $\overline{T_1} = \overline{T_2} = \overline{T_3} = \overline{T_4} = \overline{T_5}$;

Alternative hypothesis, H_1**:** $\overline{T_1} \neq \overline{T_2} \neq \overline{T_3} \neq \overline{T_4} \neq \overline{T_5}$;

Level of significance, $\alpha = 0.05$ (5%) and 0.01 (1%)

Test Statistic for comparing the treatment means,

$$F_T = \frac{TMS}{EMS} = \frac{18067.75}{2112.47} = 8.55$$

Conclusion: Calculated Value of F for comparing the treatment means, $F_T = 8.55$. Tabulated value of F at 4, 15 d.f for 0.05(5%) Level of significance, $F_{0.05} = 3.06$. Tabulated value of F at 4, 15 d.f for 0.01(1%) Level of significance, $F_{0.01} = 4.89$

As $F_T > F_{0.01}$, H_0 is rejected at 0.01 (1%) Level of Significance. F_T is considered significant at 0.01 (1%) Level of Significance Thus it is concluded that the treatment means differ significantly at 1 % level of Significance.

ANOVA Table for CRD

Sources of variation	Degrees of Freedom (d.f.)	Sum of Squares (SS)	Mean Squares (MS)	F value		
				Calculated F	Tabulated F	
					0.05 LS	0.01 LS
Treatments	$5 - 1 = 4$	$S^2_{TR} = 72271$	TMS = 18067.75	F_T = TMS/EMS = 8.55	3.06	4.89
Error	$5(4 - 1) = 15$	$S^2_E = 31687$	EMS = 2112.47			
Total	$5 \times 4 - 1 = 19$	$S^2_T = 103958$				

As F_T is significant i.e., the treatment means differ significantly, C.D. (Critical Difference) test is to be done to know actually which pair of treatment means differ significantly.

C.D. = $t_1 \times SE(d)$; t_1 – tabulated value of t for 0.05 (5%) Level of Significance at error of degrees of freedom.

SE(d) – Standard error of difference between treatment means.

$$SE(d) = \sqrt{\frac{2 \times EMS}{r}} = \sqrt{\frac{2 \times 2112.47}{4}} = 32.50$$

C.D. = 2.131 × 32.50 = 69.26

The treatment means arranged in descending order as: $\overline{T_3} > \overline{T_5} > \overline{T_1} > \overline{T_2} > \overline{T_4}$

$\overline{T_3} = (568.75)$; $\overline{T_5} = (551.75)$; $\overline{T_1} = (486.25)$; $\overline{T_2} = (430.75)$; $\overline{T_4} = (423.00)$

Difference	$\overline{T_3}-\overline{T_5}=$	$\overline{T_3}-\overline{T_1}=$	$\overline{T_3}-\overline{T_2}=$	$\overline{T_3}-\overline{T_4}=$	$\overline{T_5}-\overline{T_1}=$	$\overline{T_5}-\overline{T_2}=$	$\overline{T_5}-\overline{T_4}=$	$\overline{T_1}-\overline{T_2}=$	$\overline{T_1}-\overline{T4}=$	$\overline{T_2}-\overline{T_4}=$
	17	82.5	138.5	145.75	65.5	121.5	128.75	56	63.25	7.25
C.D. value	69.26									
S/NS	*NS*	*S*	*S*	*S*	*NS*	*S*	*S*	*NS*	*NS*	*NS*

The result of CD test is: $\underset{(ab)}{\overline{T_3}}$ $\underset{(bc)}{\overline{T_5}}$ $\underset{(cd)}{\overline{T_1}}$ $\underset{(de)}{\overline{T_2}}$ $\underset{(df)}{\overline{T_4}}$

Conclusion: The yield of Variety *C* is the highest which is at par with variety *E* but significantly higher than varieties *A, B* & *D*. Variety *E* gives second highest yield which is at par with variety *A* but significantly higher than varieties *B* & *D*. Variety *D* gives the lowest yield which is at par with Varieties *A* & *B*.

Randomised Complete Block Design (RCBD)

In this design the experimental material is divided into groups called blocks. Each block is divided into plots. Number of plots within a block is equal to no. of treatments. All the treatments are allocated in a block once and only once. Due to this reason, it is called complete block design. Since the treatments are allocated randomly in a block, the name randomized complete block design.

The blocks are formed in such a manner that the plots within blocks are as homogeneous as possible and the plots between the blocks are as heterogeneous as possible. Thus the homogeneous plots are grouped together to form a block. This is principle of local control. Thus in RCBD all the three principle of design of experiment are applied. The homogeneous sub-groups into which the experimental material is divided are called *blocks*. **Blocks are made perpendicular to the direction of variation of the experimental material.** Since in RCBD, each block is a complete replication, the words block and replication can be used synonymously. Since all the treatments are allotted in a block once and only once, so no. of plots within a block is equal to no. of treatments. Here number of blocks is equal to no. of replications.

Definition of RCBD

The technique of dividing the experimental material into '*r*' blocks and then subdividing each blocks into 't' units and allocating the treatments randomly within a block is called RCBD.

Layout of RCBD

The entire experimental material is divided into number of homogenous blocks. If there are '*t*' treatments each block is divided into '*t*' units. The units in each block are numbered from 1 to *t*. By using random number table we select '*t*' distinct random numbers from 1 to *t*. These random numbers correspond to the treatment numbers. The first selected treatment is allotted to the first unit of the block, the second treatment is allotted to the second unit of the block and so on. The randomisation is done fresh for each block in the same way.

Statistical Analysis of RCBD

The ANOVA model for RCBD is $Y_{ij} = \mu + t_i + b_j + e_{ij}$; μ = overall mean, t_i = ith treatment effect; r_j = j^{th} block/replication effect, e_{ij} = error effect.

Meaning of general mean effect: If there would have no difference in the treatments and no experimental error, then the value of the response variable for each experimental unit would have been μ.

Meaning of ith treatment effect: If there would have no block/replication effect and no experimental error (i.e., error effect), then the value of the response variable for each experimental unit of ith treatment would have been t_i. i^{th} treatment effect, $t_i = \overline{T_i} - \mu$; where $\overline{T_i}$ is the ith treatment mean

Meaning of jth block/replication effect: If there would have been no treatment effect and no experimental error (i.e., error effect), then the value of the response variable for each experimental unit of jth block/replication would have been b_j

j^{th} block/replication effect, $b_j = \overline{B_j} - \mu$; where $\overline{B_j}$ is the jth block/replication mean.

Error effect is the effect due to extraneous factors.

The Tabular Arrangement of Data

Blocks/Replications↓	Treatments→ **1 2..................i......................t**	j^{th} block total (B_j)
1	Y_{11} Y_{21}................Y_{i1} ....Y_{t1}	B_1
2	Y_{12} Y_{22}................Y_{i2} ....Y_{t2}	B_2
•	....	‘
•	....	‘
•	....	‘
j	Y_{1j} Y_{2j}..................Y_{ij}..................Y_{tj}	B_j
•	....	•
•	....	•
•	....	•
r	Y_{1r} Y_{2r}...............Y_{ir}.................. Y_{tr}	B_r
i^{th} treatment total (T_i)	T_1 T_2 ....T_i.............. ...T_t	
i^{th} treatment Mean ($\overline{T_i}$)	$\overline{T_1}\overline{T_2}$ ....$\overline{T_i}$ ....$\overline{T_t}$	

Y_{ij} = Response of jth Replication for i^{th} Treatment;

Grand total, $G = \sum_{i=1}^{t} \sum_{j=1}^{r} Y_{ij}$; Correction Factor, $CF = \frac{G^2}{r \times t}$

Calculation of sum of squares

1. Total sum of squares, Total sum of squares (TSS), $S_T^2 = \sum_{i=1}^{t} \sum_{j=1}^{r} Y_{ij}^2 - CF$

   ($\sum_{i=1}^{t} \sum_{j=1}^{r} Y_{ij}^2$ is called raw sum of squares);

2. Treatment sum of squares ($T_r.S.S$), $S_t^2 = \frac{\sum_{i=1}^{t} T_i^2}{r} - CF$; where, T_i is the ith

   treatment total i^{th} treatment total, $T_i = \sum_{j=1}^{r} Y_{ij}$

3. Block/Replication sum of squares ($B.S.S$), $S_B^2 = \frac{\sum_{j=1}^{r} B_j^2}{t} - CF$; where, B_j is

   the jth block/replication total; j^{th} block/replication total, $B_j = \sum_{i=1}^{t} Y_{ij}$

   Error sum of squares (SSE or ESS), $S_E^2 = S_T^2 - S_t^2 - S_B^2$

Calculation of mean squares→Mean square= $\frac{\text{Sum of squares}}{\text{Corresponding d.f.}}$

1. Treatment mean squares(T.M.S), $s_t^2 = \frac{S_t^2}{t-1}$;

2. Block/Replication mean squares(B.M.S), $s_B^2 = \frac{S_B^2}{r-1}$

3. Error mean square(E.M.S), $s_E^2 = \frac{S_E^2}{(r-1) \times (t-1)}$;

   Null hypothesis, $H_{0(T)}: t_1 = t_2 = \ldots = t_t$ (i.e., the treatment effects are identical)

   Or, Null hypothesis, $H_{0(T)}: \overline{T}_1 = \overline{T}_2 = \cdots = \overline{T}_t$ (i.e., the treatment means are identical)

   This is because, i^{th} treatment effect, $t_i = \overline{T}_i - \mu$; So, $t_1 - t_2 = \overline{T}_1 - \overline{T}_2$

   Alternative hypothesis, $H_{1(T)}: t_1 \neq t_2 \neq \ldots \neq t_t$ (i.e., at least any two treatment effects differ significantly)

   Or, Alternative hypothesis, $H_{1(T)}: \overline{T}_1 \neq \overline{T}_2 \neq \cdots \neq \overline{T}_t$ (i.e., at least any two treatment means differ significantly)

For block/replication

Null hypothesis, $H_{0(B)}$:$b_1 = b_2 = \ldots = b_r$ (i.e., the block/replication effects are identical).

Alternative hypothesis, $H_{1(B)}$: $b_1 \neq b_2 \neq \ldots \neq b_r$ (i.e., at least any two block/ replication effects differ significantly).

The null and alternative hypothesis can also be framed as:

Null hypothesis, $H_{0(B)}$: $\overline{B_1} = \overline{B_2} = \cdots = \overline{B_t}$ (i.e., the block/replication means are identical.

Alternative hypothesis: $H_{1(B)}$: $\overline{B_1} \neq \overline{B_2} \neq \cdots \neq \overline{B_t}$ (i.e., at least any two block/ replication means differ significantly).

This is because, j^{th} block/replication effect, $b_j = \overline{B_j} - \mu$; So, $b_1 - b_2 = \overline{B_1} - \overline{B_2}$

Level of significance, $\alpha = 0.05(5\%)$ or $0.01(1\%)$

ANOVA table for RCBD

Sources of variation	d.f.	Sum of squares	Mean squares	F
Block/ Replication	$r - 1$	S_B^2	s_B^2	$F = s_B^2 / s_E^2$
Treatment	$t - 1$	S_t^2	s_t^2	$F = s_t^2 / s_E^2$
Error	$(r - 1)(t - 1)$	S_E^2	s_E^2	
Total	$rt - 1$	S_T^2		

Test statistic, $F_B = s_B^2 / s_E^2 = BMS/EMS$

Tabulated value of F_B is obtained from F-table at $(r - 1, (r - 1)(t - 1))$ d.f. for α level of significance.

If calculated $F_B \geq$ tab F_α at $(r - 1, (r - 1)(t - 1))$ d.f. then F_B is considered significant. We thus reject H_0 and conclude that the block/replication effect (i.e., the block/replication mean) differ significantly.

If calculated $F_B <$ tab F_α at $(r - 1, (r - 1)(t - 1))$ d.f., then F_B is considered insignificant. We thus accept H_0 and conclude that the block/replication effect (i.e., the block/replication means) do not differ significantly.

Test statistic, $F_T = s_t^2 / s_E^2 = TMS/EMS$

Tabulated value of F_T is obtained from F-table at $(t - 1, (r - 1)(t - 1))$ d.f. for α level of significance.

If calculated $F_T \geq$ tab F_α at $(t - 1, (r - 1)(t - 1))$ d.f. then F_T is considered significant. We thus reject H_0 and conclude that the treatment effect (i.e., the treatment means) differ significantly.

If calculated $F_T <$ tab F_α at $(t - 1, (r - 1)(t - 1))$ d.f., then F_T is considered insignificant. We thus accept H_0 and conclude that the treatment effect (i.e., the treatment means) do not differ significantly.

If F_T is significant, we go for C.D. or L.S.D. test to find out actually which pair of treatment means differ significantly.

Critical difference(C.D.) test or L.S.D. test

Critical difference test, the C.D. test is a form of t- test. Its formula is given by, C.D.= t_α at error d.f × SE(d); where, $d \rightarrow$ difference between the treatment means tabulated t value for level of significance and error degree of freedom

$$\text{SE(d)}= SE(d) = \sqrt{\frac{2 \times EMS}{r}}$$

Two treatment means are declared significantly different at a specified level of significance if their difference exceeds the calculated CD value; otherwise they are not significantly different i.e., said to be at par. The treatment means which are at par are connected by a bar or line.

Advantages of RCBD

- The main advantage of RCBD is that it increases the precision of experiment due to the reduction of experimental error by adoption of local control.
- The amount of information got in RCBD is more efficient than CRD.
- Any no. of replication can be included in RCBD.
- Statistical analysis is simple and easy. Even when some observations are missing, the data can be analyzed by using missing plot technique

Disadvantages of RCBD

When no. of treatments is increased, the block size will increase. If the block size is large, it may be difficult to maintain homogeneity within blocks. Consequently, the experimental error will be increased. Hence, RCBD may not be suitable for large no. of treatments.

Uses

- RCBD is a versatile design. So it is the most frequently used in agricultural experiments. It is used if there is one directional variation in the experimental material.
- RCBD controls variation in one direction.

Problem

A varietal trial of paddy crop was laid out in RCBD with 5 replications for 6 varieties namely, *A*, *B*, *C*, *D*, *E* and *F*. The yield (in kg per plot) of the varieties in different replications/blocks is given below:

Varieties → Blocks ↓	*A*	*B*	*C*	*D*	*E*	*F*
1	55.7	35.6	59.2	28.3	33.3	78.4
2	45.7	39.5	58.6	44.3	44.4	89.5
3	66.7	45.8	76.9	45.7	55.5	75.3
4	46.8	50.2	80.8	55.7	66.6	78.1
5	59.0	34.5	72.6	25.6	41.4	66.8

Test whether or not the varieties differ significantly with respect to yield.

Solution

Varieties → Blocks ↓	*A*	*B*	*C*	*D*	*E*	*F*	Replication Total (R_j)	Replication Mean ($\overline{R_j}$)
1	55.7	35.6	59.2	28.3	33.3	78.4	290.5	48.42
2	45.7	39.5	58.6	44.3	44.4	89.5	322.0	53.67
3	66.7	45.8	76.9	45.7	55.5	75.3	365.9	60.98
4	46.8	50.2	80.8	55.7	66.6	78.1	378.2	63.03
5	59.0	34.5	72.6	25.6	41.4	66.8	299.9	49.98
Treatment Total (T_i)	273.9	205.6	348.1	199.6	241.2	388.1	$G = 1656.5$	
Treatment Mean ($\overline{T_i}$)	54.78	41.12	69.62	39.92	48.24	77.62		

Let y_{ij} denote the yield of ith treatment for j^{th} replication. $i = 1, 2, 3, 4, 5, 6;$ $j = 1, 2, 3, 4, 5$

Number of treatments, $t = 6$; Number of replications, $r = 5$; Number of observations, $n = r \times t = 5 \times 6 = 30$.

$$\text{Grand total, } G = \sum_{i=1}^{6}\sum_{j=1}^{5} y_{ij} = 55.7 + 35.6 + \ldots + 44.1 + 66.8 = 1656.5$$

$$\text{Correction factor, } C.F = \frac{G^2}{n} = \frac{(1656.5)^2}{30} = 91466.41$$

$$\text{Treatment Total, } T_i = \sum_{j=1}^{5} y_{ij}.. \text{ Treatment Mean , } \overline{T_i} = \frac{T_i}{5};$$

Calculation of sum of squares

Total sum of squares, $S_T^2 = \sum_{i=1}^{5}\sum_{j=1}^{4} y_{ij}^2 - CF = 99906.75 - 91466.41 = 8440.34$

Block/replication Sum of Squares, $S_B^2 = \dfrac{\sum_{j=1}^{5} R_j^2}{6} - C.F.$

$$= \frac{290.5^2 + 322^2 + \cdots + 299.9^2}{5} - 91466.41 = 92488.72 - 91466.41 = 1022.31$$

Treatment Sum of Squares,

$$S_t^2 = \frac{\sum_{i=1}^{6} T_i^2}{5} - C.F.$$

$$= \frac{273.9^2 + 205.6^2 + \cdots + 388.1^2}{5} - 91466.41 = 97421.08 - 91466.41 = 5954.67$$

Error Sum of Squares,

$$S_E^2 = S_T^2 - S_t^2 - S_B^2 = 8440.34 - 1022.31 - 5954.67 = 1463.36$$

Calculation of Mean squares : mean square $= \dfrac{SS}{df}$; SS – Sum of Squares; df – degrees of freedom

Block/Replication Mean square, RMS $= \dfrac{S_R^2}{r-1} = \dfrac{1022.31}{5-1} = 255.578$

Treatment Mean Square, TMS $= \dfrac{S_{TR}^2}{t-1} = \dfrac{5954.67}{6-1} = 1190.934$

Error Mean Square, EMS $= \dfrac{S_E^2}{(t-1)(r-1)} = \dfrac{1463.36}{(6-1)(5-1)} = 73.168$

Null hypothesis (for treatments), $H_{0(T)}$: $\overline{T_1} = \overline{T_2} = \overline{T_3} = \overline{T_4} = \overline{T_5} = \overline{T_6}$

Alternative hypothesis (for treatments), $H_{1(T)}$: $\overline{T_1} \neq \overline{T_2} \neq \overline{T_3} \neq \overline{T_4} \neq \overline{T_5} \neq \overline{T_6}$

Null hypothesis (for blocks/replications), $H_{0(R)}$: $\overline{R_1} = \overline{R_2} = \overline{R_3} = \overline{R_4} = \overline{R_5}$

Alternative hypothesis (for blocks/replications), $H_{1(R)}$: $\overline{R_1} \neq \overline{R_2} \neq \overline{R_3} \neq \overline{R_4} \neq \overline{R_5}$

Level of significance, $\alpha = 0.05$ (5%) and 0.01 (1%)

Test Statistic F for comparing the block/replications means,

$$F_R = \frac{RMS}{EMS} = \frac{255.578}{73.168} = 3.49$$

Test Statistic F for comparing the treatment means,

$$F_T = \frac{\text{TMS}}{\text{EMS}} = \frac{1190.934}{73.168} = 16.28$$

Conclusion: Calculated Value of F for comparing the block/replication means, $F_R = 3.49$.

Tabulated value of F at 4, 20 d.f for 0.05(5%) Level of significance, $F_{0.05} = 2.87$

Tabulated value of F at 4, 20 d.f for 0.01(1%) Level of significance, $F_{0.01} = 4.43$

As $F_R > F_{0.05}$, $H_{0(R)}$ is rejected at 0.05 (5%) Level of Significance. F_R is considered significant at 0.05 (5%) Level of Significance. Thus it is concluded that the block/replication means differ significantly at 5 % level of Significance.

Calculated Value of F for comparing the treatment means, $F_T = 16.28$

Tabulated value of F at 5, 20 d.f. for 0.05 (5%) Level of significance, $F_{0.05} = 2.71$

Tabulated value of F at 5, 20 d.f. for 0.01(1%) Level of significance, $F_{0.01} = 4.10$

As $F_T > F_2$, $H_{0.01}$ is rejected at 0.01 (1%) Level of Significance. F_T is considered significant at 0.01 (1%) Level of Significance. Thus it is concluded that the treatment means differ significantly at 1% Level of Significance

ANOVA Table for RCBD

Sources of variation	Degrees of Freedom (d.f.)	Sum of Squares (SS)	Mean Squares (MS)	F value		
				Calculated F	Tabulated F	
					0.05 LS	0.01 LS
Block/Replication	5 − 1 = 4	$S_R^2 = 1022.31$	RMS = 255.578	F_R = RMS/EMS = 3.49	2.87	4.43
Treatments	6 − 1 = 5	$S_{TR}^2 = 5954.67$	TMS = 1190.934	F_T = TMS/EMS = 16.28	2.71	4.10
Error	(5 − 1)(6 − 1) = 20	$S_E^2 = 1463.36$	EMS = 73.168			
Total	5 × 4 − 1 = 19	$S_T^2 = 8440.34$				

As F_T is significant i.e., the treatment means differ significantly, C.D. (Critical Difference) test is to be done to know actually which pair of treatment means differ significantly.

C.D. = $t_1 \times$ SE(d); t_1 – tabulated value of t for 0.05 (5%) Level of Significance at error of degrees of freedom.

SE(d) – Standard error of difference between treatment means.

$$SE(d) = \sqrt{\frac{2 \times EMS}{r}} = \sqrt{\frac{2 \times 73.168}{4}} = 6.045$$

C.D. = 2.086 × 6.045 = 12.62

The treatment means arranged in descending order as:

$$\overline{T}_6 > \overline{T}_3 > \overline{T}_1 > \overline{T}_5 > \overline{T}_2 > \overline{T}_4$$

$$\overline{T}_6 = (77.62);\ \overline{T}_3 = (69.62);\ \overline{T}_1 = (54.78);\ \overline{T}_5 = (48.24);$$

$$\overline{T}_2 = (41.12);\ \overline{T}_4 = (39.92)$$

Difference between the treatment means	C.D. Value	S/NS	Difference between the treatment means	C.D. Value	S/NS
$\overline{T}_6 - \overline{T}_3 = 8.00$	12.62	NS	$\overline{T}_3 - \overline{T}_4 = 29.70$	12.62	S
$\overline{T}_6 - \overline{T}_1 = 22.84$		S	$\overline{T}_1 - \overline{T}_5 = 6.54$		NS
$\overline{T}_6 - \overline{T}_5 = 28.38$		S	$\overline{T}_1 - \overline{T}_2 = 13.66$		S
$\overline{T}_6 - \overline{T}_2 = 36.50$		S	$\overline{T}_1 - \overline{T}_4 = 14.86$		S
$\overline{T}_6 - \overline{T}_4 = 37.70$		S	$\overline{T}_5 - \overline{T}_2 = 7.12$		NS
$\overline{T}_3 - \overline{T}_1 = 14.84$		S	$\overline{T}_5 - \overline{T}_4 = 8.32$		NS
$\overline{T}_3 - \overline{T}_5 = 21.38$		S	$\overline{T}_2 - \overline{T}_4 = 1.20$		NS
$\overline{T}_3 - \overline{T}_2 = 28.50$		S			

The result of CD test is : $\overline{T}_6$ $\overline{T}_3$ $\overline{T}_1$ $\overline{T}_5$ $\overline{T}_2$ $\overline{T}_4$

(ab) (be) (de) (ef) (fg) (fh)

Final Conclusion: The yield of Variety *F* is the highest which is at par with variety *C* but significantly higher than varieties *A, E, B & D.* Variety *C* gives second highest yield which is significantly higher than varieties *A, E, B & D.* Variety *A* gives third highest yield which is at par with variety E but significantly higher than varieties *B & D.* Variety D gives the lowest yield which is at par with varieties *E & B.*

Efficiency of Design *X* over Design $Y = \frac{EMS_Y}{EMS_X} \times 100$

Efficiency of one Design over another: A design would be considered if it is more efficient than the other

C.V. of the design $= \frac{\sqrt{EMS}}{\overline{X}} \times 100;$ $\overline{X}$ is the grand mean

Choice of the Design

The error d.f is always maximum in CRD. This tends to minimize the (EMS). But at the same time if heterogeneity in the experimental material is sufficiently high then this will try to increase the error sum of squares (ESS) and thus increase the error mean of squares (EMS).

Because $$\text{EMS} = \frac{\text{ESS}}{\text{Error d.f}}.$$

If we follow RCBD instead of CRD, then the error d.f in RCBD being less than that in CRD will try to increase the EMS which in turn would decrease the efficiency of RCBD as compared to CRD. But in following RCBD, the ESS is decreased as a portion of ESS is taken up by BSS (E.S.S. = T.S.S. – Tr.S.S – B.S.S in case of RCBD, whereas, E.S.S. = T.S.S. – Tr.S.S. in CRD). Thus in switching from CRD to RCBD, the decrease in error d.f. tries to increase the EMS and the decrease in ESS tries to decrease the EMS. Thus if the decrease in ESS in RCBD is substantial so that that the EMS of RCBD will be less than that of CRD in spite of decrease in error d.f. in RCBD, then use of RCBD will be beneficial.

MIX
Papier | Fördert gute Waldnutzung
FSC® C083411

Zeitfracht Medien GmbH
Ferdinand-Jühlke-Straße 7
99095 Erfurt, Deutschland
produktsicherheit@kolibri360.de